D0846709

ENVIRONMENTAL SCIENCE IN THE COASTAL ZONE:
Issues for Further Research

Proceedings of a retreat held at
the J. Erik Jonsson Woods Hole Center
Woods Hole, Massachusetts
June 25-26, 1992

Commission on Geosciences, Environment, and Resources
National Research Council

National Academy Press
Washington, D.C. 1994

R551.457   E61

MAR 21 '95

NOTICE: The project that is the subject of this report was approved by the Governing Board of the National Research Council, whose members are drawn from the councils of the National Academy of Sciences, the National Academy of Engineering, and the Institute of Medicine. The members of the committee responsible for the report were chosen for their special competencies and with regard for appropriate balance.

This report has been reviewed by a group other than the authors according to procedures approved by a Report Review Committee consisting of members of the National Academy of Sciences, the National Academy of Engineering, and the Institute of Medicine.

The National Academy of Sciences is a private, non-profit, self-perpetuating society of distinguished scholars engaged in scientific and engineering research, dedicated to the furtherance of science and technology and to their use for the general welfare. Upon the authority of the charter granted to it by the Congress in 1863, the Academy has a mandate that requires it to advise the federal government on scientific and technical matters. Dr. Bruce Alberts is president of the National Academy of Sciences.

The National Academy of Engineering was established in 1964, under the charter of the National Academy of Sciences, as a parallel organization of outstanding engineers. It is autonomous in its administration and in the selection of its members, sharing with the National Academy of Sciences the responsibility for advising the federal government. The National Academy of Engineering also sponsors engineering programs aimed at meeting national needs, encourages education and research, and recognizes the superior achievements of engineers. Dr. Robert M. White is president of the National Academy of Engineering.

The Institute of Medicine was established in 1970 by the National Academy of Sciences to secure the services of eminent members of appropriate professions in the examination of policy matters pertaining to the health of the public. The Institute acts under the responsibility given to the National Academy of Sciences by its congressional charter to be an adviser to the federal government and, upon its own initiative, to identify issues of medical care, research, and education. Dr. Kenneth I. Shine is president of the Institute of Medicine.

The National Research Council was organized by the National Academy of Sciences in 1916 to associate the broad community of science and technology with the Academy's purposes of furthering knowledge and advising the federal government. Functioning in accordance with general policies determined by the Academy, the Council has become the principal operating agency of both the National Academy of Sciences and the National Academy of Engineering in providing services to the government, the public, and the scientific and engineering communities. The Council is administered jointly by both Academies and the Institute of Medicine. Dr. Bruce Alberts and Dr. Robert M. White are chairman and vice chairman, respectively, of the National Research Council.

LIBRARY OF CONGRESS CATALOG CARD NO. 93-85862
INTERNATIONAL STANDARD BOOK NUMBER 0-309-04980-6

Additional copies of this report are available from:

National Academy Press
2101 Constitution Avenue, NW
Box 285
Washington, DC 20055
800-624-6242
202-334-3313 (in the Washington Metropolitan Area)

B-207

AFP 7408

Copyright 1994 by the National Academy of Sciences. All rights reserved.

Cover art provided by Eileen Kiliman, a native of White Plains, New York, living in Front Royal, Virginia. Earned a BBA in Marketing and a Certificate in Commercial Art from Pace University in New York. After eight years of representing graphic design firms, she has made her lifetime love of art a full time profession receiving a variety of commisions. Her style projects childhood observations combined with adult introspection.

Printed in the United States of America

5201 WOODWARD AVE.
DETROIT, MI 48202

STEERING COMMITTEE AND COASTAL WORKING GROUP

KARL K. TUREKIAN, Yale University, New Haven, CT, *Retreat Chairman*
DONALD F. BOESCH, University of Maryland, Cambridge, MD, *Working Group Chairman*[1]
ROBERT C. BEARDSLEY, Woods Hole Oceanographic Institute, Woods Hole, MA
HELEN M. INGRAM, University of Arizona, Tucson, AZ
GENE E. LIKENS, The New York Botanical Garden, Millbrook, NY
SYUKURO MANABE, NOAA Geophysical Fluid Dynamics Laboratory, Princeton, NJ
DUNCAN T. PATTEN, Arizona State University, Tempe, AZ
LARRY L. SMARR, University of Illinois, Urbana-Champaign, IL

PRINCIPAL CONTRIBUTORS

ALAN F. BLUMBERG, Hydroqual, Incorporated, Glen Rock, NJ
RICHARD F. BOPP, Rensselaer Polytechnic Institute, Troy, NY
WILLIAM M. EICHBAUM, The World Wildlife Fund, Washington, DC
DOUGLAS L. INMAN, Scripps Institution of Oceanography, La Jolla, CA
STEPHEN P. LEATHERMAN, University of Maryland, College Park, MD
RICHARD ROTUNNO, National Center for Atmospheric Research, Boulder, CO
JERRY R. SCHUBEL, The State University of New York, Stony Brook, NY
R. EUGENE TURNER, Louisiana State University, Baton Rouge, LA
JOY B. ZEDLER, San Diego State University, San Diego, CA

NRC STAFF

STEPHEN D. PARKER, Associate Executive Director, Commission on Geosciences, Environment, and Resources
GARY D. KRAUSS, Staff Officer, Water Science and Technology Board

---

[1] Donald Boesch chaired the CGER Coastal Working Group -- a follow-on activity to the retreat -- and helped to consolidate the major themes of the retreat.

COMMISSION ON GEOSCIENCES, ENVIRONMENT, AND RESOURCES

M. GORDON WOLMAN, The Johns Hopkins University, Baltimore, MD, *Chairman*
PATRICK R. ATKINS, Aluminum Company of America, Pittsburgh, PA
PETER EAGLESON, Massachusetts Institute of Technology, Cambridge, MA
EDWARD A. FRIEMAN, Scripps Institution of Oceanography, La Jolla, CA
W. BARCLAY KAMB, California Institute of Technology, Pasadena, CA
JACK E. OLIVER, Cornell University, Ithaca, NY
FRANK L. PARKER, Vanderbilt University, Nashville, TN
RAYMOND A. PRICE, Queen's University at Kingston, Ontario, Canada
THOMAS A. SCHELLING, University of Maryland, College Park, MD
LARRY L. SMARR, University of Illinois, Urbana-Champaign, IL
STEVEN M. STANLEY, The Johns Hopkins University, Baltimore, MD
VICTORIA J. TSCHINKEL, Landers and Parsons, Tallahassee, FL
WARREN WASHINGTON, National Center for Atmospheric Research, Boulder, CO
EDITH BROWN WEISS, Georgetown University Law Center, Washington, DC

Staff

STEPHEN RATTIEN, Executive Director
STEPHEN D. PARKER, Associate Executive Director
MORGAN GOPNIK, Assistant Executive Director
JEANETTE SPOON, Administrative Officer
SANDI FITZPATRICK, Administrative Associate
ROBIN ALLEN, Senior Project Assistant

Preface

On June 25-26, 1992, the National Research Council's (NRC) Commission on Geosciences, Environment, and Resources (CGER) held a retreat on "Multiple Uses of the Coastal Zone in a Changing World" at its Jonsson Study Center in Woods Hole, Massachusetts. The purpose of the retreat was to bring together a diverse group to assess the dimensions of our scientific knowledge as it applies to social and environmental issues in the coastal zone.

People have always been drawn to coastal areas for food, commerce, and recreation. Over the past half century, increases in the coastal population and accelerated land-use changes have threatened key ecosystem functions that are necessary to maintain plant and animal communities, the quality and quantity of our water supplies, and our ability to harvest the historically abundant natural resources from our bays, estuaries, and the oceans within our coastal waters.

Population and economic activity continue to concentrate in the coastal zone. The stakes riding on future management decisions affecting the coastal zone are large. On the one hand, traditional management approaches have not been adequate to halt environmental degradation and associated losses. On the other, significant economic and social costs are associated with many of the protective measures proposed, such as controlling point and nonpoint sources of pollution, imposing limits on fish harvests, or restricting coastal development. Given the complexity of the scene, there is a clear need for the scientific, managerial, and policy communities to surmount traditional barriers to support a more integrated approach to inform and guide decisions in the coastal zone.

A steering committee of commission members, working closely with CGER staff, organized and hosted the retreat. Karl Turekian acted as retreat chair. Ten papers were presented by recognized experts in the fields of ecology, oceanography, meteorology, marine biology, hydrology, geomorphology, and public policy. The preparation of the papers was carefully monitored by the steering committee. The fifty attendees participated in workshops that examined different aspects of the coastal zone--weather/climate, ocean circulation, geomorphology, rivers/estuaries, wetlands, land use, pollution, and public policy and institutional arrangements. Facilitators and scribes were selected to stimulate debate and discussion in the workshops and to record the key issues identified by each group. Following the retreat, a working group, led by Donald Boesch, helped pull together the ideas and recommendations from the retreat.

The report has two major sections: an overview and the background papers by individual authors. The overview represents a collaborative effort by the working group, commission members, and NRC staff based on a review of the papers and the major issues discussed at the retreat. The entire report has been read by a group other than the authors, but only the overview has been subjected to the report review criteria established by the National Research Council's Report Review Committee. The background papers have been reviewed for factual correctness and edited.

The Commission on Geosciences, Environment, and Resources gratefully acknowledges the generous contributions of time and expertise of the retreat participants. Special thanks are extended to those who made formal presentations, acted as facilitators to stimulate discussion, or served as scribes in the workshops. It is hoped that the discussions presented here will stimulate new ideas and research, suggest some starting points for a more holistic approach to broad resource and environmental issues in the coastal zone, and perhaps form the basis for further inquiry of specific issues by units of the NRC.

M. Gordon Wolman, Chairman
Commission on Geosciences,
Environment, and Resources

Contents

ENVIRONMENTAL SCIENCE IN THE COASTAL ZONE:
Issues for Further Research

1

Overview

BACKGROUND

The coastal zone, especially in areas adjacent to major rivers and bays, has historically attracted human settlements. It offers access to fisheries and commerce, proximity to rich agricultural lowlands, and recreational opportunities. It is no surprise that many of the world's major cities are situated along the coast. In developing nations, these cities are besieged by unprecedented population growth and overcrowded conditions. In the more developed countries, the settlement patterns, in addition to coastal cities, are characterized by nearly continuous residential and resort communities throughout the coastal areas. The coastal region of the United States saw its population nearly double, from approximately 60 million in 1940 to almost 120 million in 1980. More than half the population now lives within 50 miles of a coast, including the Great Lakes. While estimates of the rate of growth vary, it is certain that the U.S. coastal population will continue to increase in the future.

As the coastal zone becomes more developed, the demands for multiple uses continue to grow. Ever greater volumes of wastes must be disposed of safely, larger ports are needed to support increasing international trade, and there will be greater pressure for the extraction of critical mineral and energy resources located on or near the coast. Coastal areas are also more prominent in the new policies of the U.S. Navy as it shifts its focus from a global threat to regional challenges and opportunities.

In the face of more development, the requirements for a healthy environment are stronger than ever. Clean, attractive, accessible beaches are necessary to support general recreation needs as well as a tourism industry. Environmental awareness is growing, and natural beaches and green spaces are at a premium. Yet the coastal zone is made up of complex, fragile ecosystems. Accelerated land-use changes threaten key ecosystem functions necessary to maintain plant and animal communities, impact the quality and quantity of our water supplies, and reduce our ability to harvest natural resources from the coast to the limits of the Exclusive Economic Zone. The deterioration of coastal zones worldwide was a major emphasis at the historic Earth Summit (United Nations Conference on Environment and Development) held in Rio de Janeiro in June 1992. The discussion of oceans and coasts at the Earth Summit stressed both their importance in the global life support system and the

1

excellent opportunities they present for implementing sustainable development strategies. It is becoming increasingly apparent that a more coordinated approach--domestically and internationally--is needed to resolve multiple, and often conflicting, objectives and to make better use of recent scientific and technological developments.

These problems are not new. Twenty-five years ago, a report was issued by the Stratton Commission entitled "Our Nation and the Sea; A Plan for National Action" (U.S. Commission on Marine Science, Engineering, and Resources, 1969). The report cited intensive shoreline development from housing, industry, transportation, and shipping, and the potentially large impact of development on the biological productivity of the coastal region. Recognition was given to the importance of coastal and estuarine waters and marshlands in supporting seventy percent of the U.S. commercial fishing industry. The report noted the haphazard growth of federal jurisdiction in the coastal zone and the corresponding diffusion of responsibility. The result of the Commission's report was the establishment of the National Oceanic and Atmospheric Administration and the enactment of the Coastal Zone Management Act. This Act provided policy objectives for the coastal zone and authorized federal grants-in-aid to facilitate the establishment of State Coastal Zone Authorities. The Clean Water Act of 1972, and later amendments, asserted federal authority over all navigable waters, requiring uniform, minimum standards for municipal and industrial wastewater. Though improvements have been made in point source discharges, the impacts of other sources of pollution have become ever more apparent. In spite of significant advances, these problems remain with us because they are difficult to resolve, both scientifically and socially.

The coastal zone is a variable and often unpredictable environment, influenced by inland environments (e.g., land use, runoff, and groundwater outflow), by the atmosphere, and by the ocean. Consequently, the systematic integration of many fields of science is required to understand these different influences and to develop a comprehensive strategy for dealing with natural and anthropogenic influences on the coastal zone. To assess the dimensions of our scientific knowledge as it applies to societal problems in the coastal zone, the National Research Council's (NRC) Commission on Geosciences, Environment, and Resources (CGER) held a retreat on "Multiple Uses of the Coastal Zone in a Changing World" on June 25-26, 1992, at its Jonsson Study Center in Woods Hole, Massachusetts. The meeting brought together approximately 50 experts with science, technology, and policy perspectives to explore how multidisciplin- ary efforts could effectively address environmental policy concerns in the coastal zone. The purpose of the meeting was to consider current approaches to research and management of the coastal zone and, in the process, to identify overlaps, gaps, and opportunities for synthesis. The participants represented both the research community and the major federal programs concerned with coastal issues. Ten working papers were presented and discussed at the meeting. They were followed by workshops that examined different aspects of the coastal zone--weather/climate, ocean circulation, geomorphology, rivers/estuaries, wetlands, land use, pollution, and public policy and institutional arrangements--and identified key issues confronting science and management.

This overview briefly summarizes the findings of the group with regard to coastal environmental concerns. At the end of the overview, some issues are presented that may merit the future attention of a larger audience, especially the federal agencies involved with coastal science and policy and the various boards and committees within CGER that advise these agencies. It is hoped that the ideas presented here will suggest some starting points for the collaboration of these agencies and the NRC on the formulation of a more holistic approach to broad resource and environmental issues.

SUMMARY OF PAPERS PRESENTED

The papers presented at the retreat and reproduced here cover a spectrum of topics, including the state of coastal research, specific modeling needs, analysis and management of pollution problems, and the status of research funding. The papers can be recapped with respect to their major points as follows:

A Synopsis of Coastal Meteorology: A Review of the State of the Science, Richard Rotunno. In a review of the state of the science in coastal meteorology, Dr. Rotunno points out that we have the tech-nical capacity--based on greater scientific understanding as well as improvements in computational skills--to develop comprehensive and credible atmospheric and ocean circulation models. Such models would improve our ability to predict a storm's development and intensity, with the potential benefit of earlier warning and better preparation of emergency plans.

Modeling Transport Processes in the Coastal Ocean, Alan F. Blumberg, Richard P. Signell, and Harry L. Jenter. The authors discuss the processes governing the transport of water, dissolved substances, and particles in a three-dimensional circulation model of Massachusetts Bay. Detailed models that are now possible can be extraordinarily helpful for making decisions about the assimilation capacity for various classes of pollutants, optimal siting of outfalls, the impacts of jetties, and other coastal zone projects.

Coastal Geomorphology, Stephen P. Leatherman, A. Todd Davison, and Robert J. Nicholls. The authors present the issues surrounding the various methods of protecting beaches from erosion. In spite of large investments nationwide, there is a critical lack of research and post-project monitoring that could provide a better understanding of the performance and maintenance characteristics of different beach protection options.

Rivers and Estuaries: A Hudson Perspective, Richard F. Bopp and Daniel C. Walsh. In a discussion of the contamination of the Hudson River and its estuary, the authors point out the significant and largely untapped potential for collaboration between research scientists and agencies involved in monitoring programs -- a collaboration that could make these programs serve both research and regulatory objectives better and at lower cost.

Types of Coastal Zones: Similarities and Differences, Douglas L. Inman. In this paper, the tectonic origins and processes shaping coastal regions worldwide are compared and contrasted by Dr. Inman. The configuration of the continental shelf, coastal climate, and local exposure to waves, wind, tides, and currents are among the many factors that affect the stability of beaches, their capacity to protect the littoral plant and animal communities, and the ability of beaches to withstand human activities. This knowledge is essential in defining sustainable coastal development.

Landscapes and the Coastal Zone, R. Eugene Turner. Dr. Turner presents a "landscape perspective" on coastal management and research. He suggests an approach that can be used to examine environmental problems in a regional context including everything from small watersheds to major river basins. A landscape perspective can be combined with quantitative methodologies to help tackle difficult, multi-dimensional management problems such as eutrophication of coastal waters.

Coastal Wetlands: Multiple Management Problems in Southern California, Joy Zedler. Dr. Zedler provides insight into the distinctive characteristics of Pacific Coast wetlands and the problems that can arise from policies that encourage the application of general environmental standards to specific regional situations. She points out how coastal wetlands are threatened by multiple-use management strategies, and how the implementation of certain mitigation policies can cause unforeseen and unwanted consequences to Pacific Coast wetlands.

Coastal Pollution and Waste Management, Jerry Schubel. Dr. Schubel reviews recent assessments of the major pollution problems facing the coastal zones throughout the world, and suggests that without a new paradigm that recognizes the need for a coordinated, multi-disciplinary research program in coastal sciences, the prospect for improvements in coastal ocean quality is bleak.

Coastal Management and Policy, William Eichbaum. Dr. Eichbaum discusses how the translation of science into sound coastal management policy is hindered by the public's difficulty in understanding an inherently complex, and often invisible, coastal and marine environment. A management approach is proposed for identifying and dealing with the most important threats to the quality of the coastal zone and for providing a framework for appropriate, sustainable development.

Research and Development Funding for Coastal Science and Management in the United States, Richard Turner and Jerry Schubel. The authors examine the pattern of research funding and trends in the coastal science and engineering professions. The current research climate is analyzed, and recommendations are made for achieving better interactions among the government, education, and industry sectors in their support of long-term research programs.

PROBLEMS WITH TRADITIONAL APPROACHES TO COASTAL MANAGEMENT

The participants at the CGER retreat discussed the complexity of many coastal issues. Some of the following examples illustrate why traditional management approaches have failed to reach workable and comprehensive solutions.

The Federal Water Pollution Control Act of 1972, and the subsequent amendments to it embodied in the Clean Water Act, set the nation on a fairly rigid course of environmental protection based on uniform, minimum federal standards for municipal and industrial wastewater treatment. The nation's water quality has improved significantly, due largely to improvements in the quality of point source discharges. However, nonpoint sources of pollution, such as urban and agricultural runoff and atmospheric input, still cause severe water quality problems. While the need for improvement of estuarine and coastal waters quality has been acknowledged, and more attention is being paid to the control of nonpoint sources of pollution (e.g., the 1990 amendments to the Coastal Zone Management

Act), the current approach to managing coastal waters is still based on the application of uniform federal standards to point sources. This approach does not always lead to improved water quality that meets federal and regional objectives. Regional differences in coastal ecosystems also can lead to different regional priorities. For example, in the highly developed estuaries of the Atlantic and Gulf coasts, excess nutrients are causing eutrophication and damage to fisheries. On the Pacific coast, however, the quick-flowing rivers and ocean currents and relatively deeper, more narrow continental shelf tend to dissipate nutrient loadings. Other problems, such as habitat loss from over-development or near-shore contamination from stormwater runoff, may have a higher priority in those areas.

Human alterations of the landscape have caused large-scale hydrologic changes. As a result of massive public works programs to improve navigation and to control floods, the Mississippi River has been shortened by 229 km and is extensively diked and leveed. The monumental efforts required to keep the lower Mississippi River from its natural tendency to meander have become the focus of popular writings by John McPhee (1989) and others. The summer of 1993 saw massive flooding of the upper Mississippi valley, demonstrating the sensitivity of major stretches of the river to rare hydrologic events. While intensive engineering of the river system protects local communities during normal high-water periods, it can also magnify the force and height of extreme flood waters with costly and tragic results. In his paper, Eugene Turner explains that as a result of the massive alterations to the Mississippi River system, its historical capacity to absorb phosphorus and nitrogen runoff as well as wastewater discharges has been diminished. Excess nutrients are delivered downstream in large "pulses", altering the natural Chemical balance in estuaries and bays, and changing the composition of the marine phytoplankton community in the Gulf of Mexico. One result is that diatoms--an important food for marine fish and invertebrates--are less abundant, and other types of less desirable algae, such as flagellates, have been increasing.

Large changes to the landscape are also being made by society through the accumulated impact of small-scale decisions. For example, the issuance of many small dredge-and-fill permits--each one decided without regard to cumulative impact--has resulted in significant wetland losses in coastal Louisiana. In developed areas across the nation small creeks have been controlled by cementing or rock-filling channel banks, often increasing erosion, siltation, and flooding downstream and destroying critical marsh habitat.

One of the most significant problems for California's southern coastal areas is the disturbance and loss of coastal wetlands. Ironically, in a region of water scarcity, coastal wetlands are threatened by too much fresh water. In her paper, Joy Zedler explains how intermittent streams, which normally provide seasonal fresh water to coastal lagoons, become year-round rivers due to a constant flow of treated municipal wastewater discharge. The dilution of salt marshes and loss of endangered species habitat are direct results. Additionally, highway construction and urban development tend to close off estuary inlets, effectively reducing the natural tidal flushing and further increasing the volume of fresh water. Current environmental mitigation policies seek to compensate for a wetland loss by restoring a comparable wetland elsewhere. However, so much wetland acreage has been appropriated already that trying to mitigate losses on an acre-for-acre basis is extremely difficult. This difficulty is illustrated in the San Diego Bay mitigation project where a disturbed upper inter-tidal marsh was converted to a cordgrass marsh to help encourage the establishment of a federally endangered bird--the light-footed clapper rail. Unfortunately, the strict habitat requirements of the endangered rail were not adequately met (nutrient content and cordgrass height, among other factors, were insufficient), and evidence indicates that the site may not accomplish its intended purpose. To make matters worse, taking out the original high marsh caused inadvertent habitat loss to a different bird--the Belding's savannah sparrow--one that is on the state list of endangered species.

Coastal communities are particularly vulnerable to the natural hazards of flooding, winds, and erosion. The communities of southern Florida have still not recovered from the devastation of Hurricane Andrew, with insured losses estimated at 20 billion dollars. About 200,000 people died in Bangladesh in April 1991 from the severe flooding accompanying a cyclone, and 10 million were left homeless according to a recent United Nations report (United Nations Center for Regional Development, 1991). While the United Nations cites reductions in the death toll worldwide compared to previous years, the tragic consequences of coastal storms are often extreme. Especially in developing countries, the impacts of natural disasters on coastal areas are exacerbated by: environmental change (e.g., denuding of forest cover); a weak or nonexistent infrastructure for sanitation, roads and buildings; an increasingly urbanized population; and the lack of alternative living sites for the disenfranchised people who often live in low-lying, flood-prone areas. Of course, in developed countries people also occupy (and re-occupy) scenic but risky areas that may be flood-prone, unstable, or subject to high winds. The societal cost of repairing damages and resettling communities is often not considered until after the tragic events.

As pointed out by Stephen Leatherman in his paper, 90 percent of U.S. beaches and 70 percent of beaches worldwide are eroding. While some of this beach material is recaptured on other beaches, there has been a net loss of coastal sediments--due, in part, to an incremental sea level rise worldwide, and in part, to human activities. Along the Pacific coast, rivers and streams are important sources of sediment to beaches. The building of dams for flood control, debris control, hydroelectric power, and water supply has reduced the supply of sand available to the coast by interfering with natural flooding and sediment transport processes.

For Atlantic coast beaches, longshore currents along barrier islands are a major influence. Unfortunately, many efforts at beach protection have had negative effects on neighboring beaches. For example, Ocean City, Maryland, is built on a barrier island that was breached by a major hurricane in 1933. In response, the city built a long jetty on the south end of the beach to catch and hold the net longshore movement of sediment from the north. The impact on the adjacent shoreline has been substantial; the immediate shoreline to the south has broken away and migrated landward the complete width of the barrier island in the past 50 years. Eventually, the northern end of Assateague Island, located to the south of Ocean City, will merge with the mainland coast, with a consequent loss of valuable habitat and coastal protection from storms. While millions of dollars are spent on various forms of beach protection and replenishment, understanding of the performance and impacts of the various approaches is still poor because of a lack of effective monitoring and research programs. There is also little research on the public policy issues associated with beach protection and replenishment (e.g., who bears the costs and who benefits?).

The richness and diversity of environments in the coastal zone present a significant challenge to scientific predictive modeling. Compared to physical processes in the open ocean and those further inland, the coastal system is quite variable. Local tides, currents, and topographic factors combine with larger-scale ocean circulation and weather patterns to create complex interactions. In his paper, Richard Rotunno notes that our skill in predicting the weather and in modeling contaminant transport processes is generally poor for the coastal zone compared to other areas, and information about long-term climate characteristics in the coastal zone is likewise lacking. Predictive modeling for oceanic and atmospheric circulation is limited by our understanding of the controlling physical processes such as mixed layer dynamics, and forcing and feedback mechanisms, and by our ability to incorporate real observations into the models effectively. With the advent of new observing systems and the computing power to accommodate finer scale models, there is an opportunity to develop a better understanding of many of these processes.

Even with better scientific understanding and predictive capabilities, there remains a gap in the capacity of government to manage coastal resources. Policy formulation in the coastal zone is complicated by the fragmentation of governance among local, state, and federal authorities. While scientific consensus can often influence the public debate, the science of the coastal environment is extraordinarily complex, and our understanding of the nature and causes of environmental problems may be tentative and inconclusive. This situation poses serious problems for policy-makers trying to convey information about environmental issues to the public. Many severe coastal problems are not easily brought to the public's attention, and neither managers nor citizens can be expected to have an intuitive sense of their urgency. In his paper, William Eichbaum illustrates this problem with two examples. Between 1978 and 1980, over 80 percent of the submerged aquatic vegetation in the Chesapeake Bay died, and scarcely anyone noticed. (These grasses, an important food source and habitat for much of the aquatic life in the Bay, have shown a steady increase since then.) However, one can imagine the public reaction if 80 percent of the forest resources of the Bay's watershed had died during the same period. Conversely, the public's perception of a problem can often be out of proportion to the real level of risk. During the summer of 1988 when a relatively small number of syringes washed up on the beaches of New Jersey, a huge public outcry resulted. The public's response was not based on scientific assessments and may have obscured more serious pollution problems. An essential challenge for all scientists is to organize and present scientific information in a manner understandable to policy-makers and to the general public.

INTEGRATED COASTAL MANAGEMENT

In the United States, many agencies are now implementing various integrated approaches to protect, conserve, or restore coastal environments. Examples include the EPA National Estuary Program and interagency programs for the Great Lakes, the Chesapeake Bay, and the Gulf of Maine. The effort to consider the relative influences and pollution sources from the atmosphere, the watershed, the estuary, and the continental shelf, as well as their interconnections is a daunting challenge to federal, state, and local planners, and to the scientific and technical community. In addition to technical complexities, this approach poses the challenge of mediating among the different sectors of society vying for use of the coastal zone, and of integrating responsibilities among the different levels of government with roles in managing the coastal zone.

Scientists are being asked to consider difficult questions such as: "What effect do auto emissions have on nutrient enrichment and oxygen depletion in coastal waters?" The stakes riding on the management decisions are large. On the one hand, traditional management approaches are not adequate to halt environmental degradation and associated losses. On the other hand, there are significant economic costs associated with many of the protective measures proposed, such as controlling point and nonpoint sources of pollution, imposing limits on fish harvests, and restricting coastal development. Consequently, there is a clear need for the scientific community to surmount traditional barriers between disciplines, to more actively involve social scientists (law, policy, planning, and economics), and to communicate the results of their more comprehensive analyses to environmental policy-makers and managers. There is an equally important need to examine the existing governmental structures and to define alternatives that will address coastal management issues in a more integrated way.

One of the major recommendations arising from UNCED was that national management of coasts and oceans (including the Exclusive Economic Zones) should be "integrated in content and precautionary in ambit." In addressing the problem of coastal pollution, a recent NRC report, *Managing Wastewater in Coastal Urban Areas* (1993), defines integrated coastal management as an "ecologically based, iterative process for identifying, at a regional scale, environmental objectives and cost-effective strategies for achieving them." Implicit in the concept of integrated coastal management is the infusion of the best scientific knowledge available into the design of research, monitoring, and management programs, and assistance in the development of new methods and techniques for better management of coastal resources. The dimensions of integrated coastal management, however, go beyond the need to understand the science and management of the coastal zone ecosystems. The broad social questions dealing with human settlement patterns, the dynamics of development of various resources and uses, and the decision making structures for governance of the coastal zone are equally important considerations that should be explored in concert with progress in the fields of environmental science and technology. Ultimately, an integrated, decision-making framework would support appropriate and sustainable development of human uses in the coastal zone.

SUGGESTED ISSUES FOR THE NATIONAL RESEARCH COUNCIL AND ITS CONSTITUENCIES TO PURSUE

The Woods Hole retreat summarized here addressed a broad array of subjects, including shoreline processes, eutrophication, monitoring, modeling, the use of scientific information in environmental management, and the adequacy of present research endeavors. Many of the issues raised by the retreat participants have been at least partially addressed in recent or ongoing NRC studies. In addition to *Managing Wastewater in Coastal Urban Areas* mentioned above, some other relevant NRC reports are *Managing Coastal Erosion, 1990*; *Managing Troubled Waters: The Role of Marine Environmental Monitoring, 1990*; *Working Together in the EEZ: Final Report of the Committee on*

Exclusive Economic Zone Information Needs, 1992; *Coastal Meteorology - A Review of the State of Science, 1992*; *Oceanography in the Next Decade, 1992*; and *Restoration of Aquatic Ecosystems, 1992*. Other important topics are slated to be addressed by studies now in the planning stage or by follow-up activities; these include response to sea-level rise, contaminated sediments, new perspectives on watershed management, marine biodiversity, eutrophication, wetlands characterization, and marine environmental monitoring.

With this in mind, the following issues are highlighted as significant priorities that should be examined closely by the relevant NRC boards and the federal agencies they serve.

Modeling Needs for Coastal Management

While traditional approaches to environmental management of coastal areas tend to focus on the coastal fringe, an integrated approach would be regional in scope, extending the geographic area and the types of environments that need to be managed farther inland, often to an entire watershed, as well as offshore. Fate-and-transport models are increasingly significant management tools for assessing air and water pollution levels. However, for these models to be effective within the framework of integrated coastal management, they must account not only for the near coast environment, but also for processes within the watersheds feeding the coast, synoptic-scale atmospheric events, and ocean circulation.

Models in use for coastal systems generally fall into one of two categories: physical circulation models (looking at currents, tides, and wind) or those based upon "ecosystem" processes, either biological or geochemical (such as carbon, energy flow, nutrient cycling, or population dynamics). The physical models need to be improved through a better understanding of circulation processes such as mixed layer dynamics, and feedback mechanisms between the ocean and atmosphere, and better utilization of observational data. To be useful, process coupling is needed to link both ecosystem and physical models. The most advanced water-quality models for dissolved oxygen and nutrients are approaching the goal of process coupling. These models are fundamentally based on circulation, but are increasingly sophisticated in their incorporation of biologically and geochemically mediated processes and should be developed further. Efforts to improve the models are still limited by data and fundamental understanding of processes. Until very recently, modeling efforts have also been limited by computer capability. Now, with access to expanded computational power, some of these new approaches may be developed. Also, new observation systems have increased the range and resolution of measurements, permitting researchers to see more complex weather, land, and ocean systems more clearly.

A possible next step would be a workshop or other forum at which the future directions for coastal predictive modeling can be discussed. One aim for such a forum would be collaboration on field experiments and modeling to improve our understanding of the dynamics of coastal atmospheric and ocean circulation, and the validation of high-resolution, nested-grid mesoscale forecasting models. The forum would also build upon the recently initiated, interdisciplinary research program by the National Science Foundation on Coastal Ocean Processes. Other discussions might address such questions as how current monitoring systems can be improved, and how models can make use of observational (including real-time) data more efficiently. Recognizing the importance of communicating scientific information (including better estimates of experimental and modeling uncertainties) to the public, participants might explore whether large scale modeling can become a tool to integrate the requisite social and economic variables of importance to policy-makers.

Improvements in Coastal Monitoring and Data Archiving

A key impediment to model development is the current inadequacy of available data for validating these models. There seems to be a significant and largely untapped potential for collaboration across disciplinary studies between research scientists and the agencies involved in monitoring programs. Data are often collected inconsistently, by different groups and for different objectives. Information retrieval is frequently difficult, and comparisons are questionable. While new systems and enhancements of monitoring networks are certainly needed, there is much to be gained by a closer linking of existing land, air, and sea monitoring activities with research objectives. For example, most state departments of environmental protection collect water-column and sediment-monitoring data as part of their regulatory requirements. Collaboration with research scientists could make these data-collection efforts also serve local and regional research programs, which in turn would ultimately provide better insights and techniques to the regulatory agencies.

For modeling programs to be effective, information must be easily retrievable and observational monitoring should be designed to provide adequate data for forecast models. Existing national repositories of data, such as U.S. EPA's STORET for the archiving and analysis of water-quality monitoring programs, and information networks, such as the U.S. Geologic Survey's NAWDEX (the National Water Data Exchange--a catalog of federal and non-federal water programs), are extremely valuable, but not complete. Satellite and airborne observing systems (e.g., Thematic Mapper, SPOT, NEXRAD, ASOS and Wind Profilers) have increased their range and resolution to see earth features and complex weather systems more clearly, and the computer power to accommodate finer grid scales is now available. Obtaining the necessary data for a comprehensive, regional assessment, however, will continue to be an expensive undertaking even when various groups have already collected relevant information. The new Title 4 amendment to the Marine Protection, Research, and Sanctuaries Act is an important step in helping to establish regional marine research and monitoring programs. Still, there is much to be learned on how best to coordinate local, state, and federal data collection to enhance the return on monitoring investments.

A useful approach to the problem of incomplete or inadequate data would be to conduct a comprehensive assessment of the current monitoring activities that affect coastal environments, and to examine options for and benefits of better coordination of efforts to provide adequate data for forecast models.

Efficacy of Environmental Trading and Credits
for Mitigating Coastal Habitat Degradation

Regulations that govern development in critical coastal habitats such as wetlands, frequently require preservation or restoration of an equal number of acres of like habitat to offset the loss from development. While the practice of land "swaps" or set-asides tied to development rights often has merit, the premise of habitat equivalence (i.e., "take an acre here, give an acre there") is problematic because of scale effects and the relationship of the habitats to those in the surrounding landscape. Just as the economic values of some coastal areas are greater than others for development purposes, it is increasingly clear that the environmental values of coastal habitats vary within a given coastal region. A recent NRC report, *Restoration of Aquatic Ecosystems (1992)*, noted that wetland restoration activities that are driven by regulatory requirements are fraught with poor design, involve land selected without consideration of landscape-level study and are rarely followed up. It stands to reason that protection or restoration can be accomplished more effectively -- and potentially at less cost -- by strategic targeting of important habitats and better valuation of natural resources in a regional context.

The merits of various regulatory approaches involving environmental trading should be critically examined in the context of integrated coastal management, the efficacy of their applications, broader economic considerations, and the research needed to improve the soundness of environmental trading in its many forms. Such an examination ultimately could provide guidance for using environmental credits to manage impacts. Both commercial interests and the environment would benefit from a better definition of the conditions and constraints appropriate for land swaps.

Institutional Arrangements

The institutional relationships that characterize governance of the many activities along the coast are complex, fragmented, and compartmentalized. Federal statutes (e.g., the Clean Water Act; the Coastal Zone Management Act; the Oil Pollution Act; the Resource Conservation and Recovery Act; the Comprehensive Environmental Response, Compensation, and Liability Act; the Marine Protection, Research and Sanctuaries Act, the National Environmental Policy Act; the Outer Continental Shelf Lands Act; and the Coastal Barrier Resources Act) and other laws dealing with maritime pollution and land use constitute a framework implemented by a variety of agencies at the federal, state, and local levels. These overlapping yet incomplete responsibilities for natural resource management, land use planning, water quality protection, and other activities have led to a fragmentation of outlook and decision-making. If our problems in the coastal zone are to be addressed effectively, there must be better integration of the planning, resource management, and water-pollution control functions. Regional bodies for water quality management were instituted with some success in the early 1970's. At the time, however, a federal emphasis on construction overwhelmed most of these efforts. Current efforts by the EPA and coalitions of state and local agencies toward regional management are promising.

The NRC report *Managing Wastewater in Coastal Urban Areas* recommends that a process of integrated coastal management be instituted for coastal areas in which existing Clean Water Act regulations are not achieving regional objectives.   That report finds that existing institutional arrangements should be better coordinated and that in some places new, regional institutions may be required to realize the promise of integrated coastal management.   It notes however that while a centralized, regional agency, in concept, appears to be an attractive institution for integrated coastal management, it may not in fact be the best alternative.   More experience needs to be gained and an assessment made of the effectiveness of various institutional arrangements.   While the issues covered at the CGER Woods Hole Retreat were broader in focus than those covered in *Managing Wastewater in Coastal Urban Areas*, the next steps suggested here are compatible with those suggested in that report.

Advancing these concepts to implementation will require the collaborative action of the scientific and resource management communities, and the concerned public.   The NRC may find it useful to serve as a convener to bring together the EPA, NOAA, and other relevant agencies and parties to begin to formulate a plan that will foster the growth of this new paradigm.

In the face of ever increasing demands for coastal development, resource managers must be prepared with options for maintaining and restoring vital natural resources.   The ideas posed in this overview and in the papers that follow suggest the types of collaboration and studies that would lead to a better integration of society's growth and environmental objectives in the coastal zone.

REFERENCES

Mcphee, J. 1989. The Control of Nature. New York: Noonday Press.

NRC (National Research Council). 1990. Managing Coastal Erosion. Washington, D.C.: National Academy Press.

NRC (National Research Council). 1990. Managing Troubled Waters: The Role of Marine Environmental Monitoring. Washington, D.C.: National Academy Press.

NRC (National Research Council). 1992. Working Together in the EEZ: Final Report of the Committee on Exclusive Economic Zone Information Needs. Washington, D.C.: National Academy Press.

NRC (National Research Council). 1992. Coastal Meteorology - A Review of the State of Science. Washington, D.C.: National Academy Press.

NRC (National Research Council). 1992. Oceanography in the Next Decade. Washington, D.C.: National Academy Press.

NRC (National Research Council). 1992. Restoration of Aquatic Ecosystems. Washington, D.C.: National Academy Press.

NRC (National Research Council). 1993. Managing Wastewater in Coastal Urban Areas. Washington, D.C.: National Academy Press.

United Nations Center for Regional Development. 1991. Cyclone Damage in Bangladesh. UNCRD, Nagoya, Japan.

U.S. Commission on Marine Science, Engineering, and Resources. 1969. Our Nation and the Sea; A Plan for National Action. Washington, D.C.: U.S. Government Printing Office.

2

A Synopsis of Coastal Meteorology:
A Review of the State of the Science

Richard Rotunno
National Center for Atmospheric Research

INTRODUCTION

According to a recent study by the Department of Commerce, almost half the U.S. population lives in coastal areas and so is affected by the unique weather and climate of the coastal zone. Under the auspices of The National Academy of Sciences, the Panel on Coastal Meteorology has just completed a study of the state of the science of coastal meteorology. This presentation will cover the highlights of the study by concentrating on the perceived major scientific problems and the opportunities for progress.

Coastal meteorology is the study of meteorological phenomena in the coastal zone caused, or significantly affected, by the sharp changes that occur between land and sea in surface transfer or elevation. The coastal zone is subjectively defined as extending approximately 100 km to either side of the coastline. Examples of coastal meteorological phenomena include the sea breeze, sea-breeze-related thunderstorms, coastal fronts, marine stratus, fog and haze, enhanced winter snow storms, and strong winds associated with coastal orography. Increased knowledge of several or all of these is important for studies on the physical and chemical oceanography of the coastal ocean. The practical application of this knowledge is vital for more accurate prediction of the coastal weather and sea state, which affect defense, transportation and commerce, and pollutant dispersal.

The dynamical meteorology of the coastal zone may be thought of in terms of three subsidiary ideal problems; these three problems formed the organizational basis of our study. The first problem is one in which the coastal atmospheric circulation is primarily driven by the contrast in heating, and modulated by the contrast in surface friction, between land and sea. The second problem is one in which the primary influence is due to the steep coastal mountains whose presence may induce strong along-shore winds and other complex tow patterns. The third class of phenomena broadly consists of larger-scale meteorological systems that, by virtue of their passage across the coastline, produce distinct smaller-scale systems. Of course, reality is always some combination of these idealized problems.

THE ATMOSPHERIC BOUNDARY LEVEL

The transfer of heat, momentum, and water vapor between the atmosphere and the lower surface (be it land or sea) is basic to these three ideal problems. As such, our study begins with an assessment of, and prospects for improvement in, our understanding of the approximately 1 km-deep layer of air adjacent to the surface called the atmospheric boundary layer (ABL). Study of the ABL is intended to reveal how the effects of surface transfers are distributed upward. The model of the ABL is best understood when it is cloud-free, convective, and horizontally homogeneous. However, near the coast, the ABL is anything but. Stratus, fog, and drizzle complicate the situation, as they depend on a complex interplay between cloud physics, radiation, and turbulence. Perhaps the most severe scientific problem is how to treat boundary layers that are not horizontally homogeneous. Over land, there is still significant uncertainty concerning the nature of surface transfer from terrain with variation in vegetation and usage, such as occurs along the coast. Over the ocean, those surface transfers are determined by the sea state, which in turn is determined by the atmospheric flow, which is influenced by the surface transfers, etc. This fundamental coupling has long been recognized. However, there is another order of complexity over the coastal ocean, because there the sea state is significantly influenced by the ocean shelf.

Areas in need of research are

• ABL processes in inhomogeneous and nonequilibrium conditions (better understanding of these may lead to better surface-flux and mixed-layer scaling theories);

• fundamental relationships between the ocean wave spectrum, the surface fluxes, and bulk ABL properties; and

• coastal marine stratocumulus.

THERMALLY DRIVEN EFFECTS

Although the recognition of the land-sea breeze dates back to antiquity, the deeper understanding needed to make accurate forecasts is still lacking. The land–sea breeze is produced by the generally different temperatures of the land and sea, which produce an across-coast, air-temperature (density) difference. After this circulation begins, however, it modifies the conditions that produced it; the difficulty in making precise predictions lies in the difficulty with understanding more precisely the nature of this feedback. The aforementioned uncertainties in our understanding of the ABL are certainly central problems here. Beyond the simple two-dimensional picture, coastline curvature, near-shore islands, and different synoptic-scale wind orientation present important scientific problems. Perhaps the most challenging problem is the interaction of the land–sea breeze with cumulus convection. Issues associated with two special types of thermally driven phenomena (coastal fronts and ice-edge boundaries) are also discussed in the study.

Areas identified for further study are

- observational and modeling studies of the land–sea breeze to cover the entire diurnal cycle, with emphasis on improving knowledge of offshore regions;
- the fine-scale structure of the sea-breeze front, including the associated vertical motions, and internal boundary layers above complex coastlines and heterogeneous surfaces;
- three-dimensional interactions of the land-sea breeze with variable synoptic-scale flow, nonuniform land and water surfaces, irregular coastlines, and complex terrain;
- dynamical interactions of the land–sea breeze with stratus clouds and with precipitating and nonprecipitating cumulus convection;
- geographical distribution, spatial coverage, and modes of propagation of coastal fronts; and
- processes of heat and moisture flux from leads and polynyas.

THE INFLUENCE OF OROGRAPHY

Coastal mountain ranges can significantly affect coastal meteorology. In many situations, the coastal mountains act as a barrier to the stably stratified marine air. Thus air with a component of motion toward the barrier must eventually turn and flow along the barrier. Also, the coastal mountains may act like the side of a basin within which the marine air is contained; under the influence of the earth's rotation, waves known as *Kelvin waves* may propagate along the basin-wall-like coastal mountain. Special bounder-layer flows are also observed to be under the influence of the coastal mountains. For example, during the Coastal Ocean Dynamics Experiment, a strong along-shore jet was documented. It had a strong diurnal component as evidenced by the depression on the marine inversion near the coastal mountains during the day. The boundary-layer structure showed interesting complexity inasmuch as the potential temperature was well-mixed to the inversion but the wind speed increased strongly through the same layer. Phenomena that appear similar to flow separation in classical fluid dynamics also occur in the lee of capes and other coastline salients. These types of motion are important components of the meteorological problem in these coastal regions.

Areas identified for further study are

- case studies of structure and path of storm systems modified by coastal orography;
- climatology of synoptic regimes conducive to coastally trapped phenomena;
- methods to include coastal phenomena in numerical forecast models.

INTERACTIONS WITH LARGER-SCALE SYSTEMS

As larger-scale meteorological systems move across the coast, they are affected by some combination of the effects discussed in the previous two paragraphs; in some situations, distinct subsystems, which would not exist without the coastal influence, are produced. Examples of these effects include cyclogenesis enhanced at the east coast of the United States as upper-level disturbances cross the Appalachians and encounter the strong baroclinic zone at the coast; flow along the coast in winter with strong cooling of the air on the landward side, leading to the formation of fronts; and land-falling hurricanes whose low-level flows are modified so as to favor the formation of tornadoes.

Areas identified for further study are

• dynamics of the local intensification of cyclone winds by coastal topography and the resulting modification of storm intensity and motion;

• the cause of tornadoes associated with land-falling hurricanes;

• the influence of the coastal-heating discontinuity in the along-shore propagation and local intensification of coastal fronts; and

• the influence of coastal fronts on midlatitude coastal cyclogenesis.

INFLUENCES ON THE COAST OCEAN

In general, the ocean affects, and is affected by, the atmosphere. We discuss, next, aspects of this interaction that are particularly important for the coastal zone (shelf waters). In the northern hemisphere, an along-coast wind with the coast on the left brings the sea into motion in the along-coast direction. Due to the Coriolis effect, the water motion is deflected away from the coast necessitating its replacement by water from below—this phenomenon is know as coastal upwelling. The water from below is colder and, in general, is of different chemical and biological composition. The details of the cross-shelf transport (necessary to feed the upwelling) are poorly understood, since the ocean is responding to atmospheric influences over a large range of time and space. This wind-stress data from the Coastal Ocean Dynamics Experiment shows the mean and a considerable standard deviation. Also, the along-shore ocean currents may be highly irregular. There is evidence that some of the irregularity is due to wind-stress variations along and across the coastal zone.

The colder water now along the coast means there is yet another across-coast temperature difference that can produce changes in the atmospheric circulation, which can affect the ocean, etc. Interactions of this nature are important to the understanding of the coastal ocean and the chemical and biological processes occurring there.

Areas identified for further study are

• the coupled ocean-atmosphere processes that control the interactions between the wind field, ABL structure, and upper ocean;

• the local physical and chemical processes governing air–sea fluxes of momentum, heat, moisture, particulate, and gas within an inhomogeneous coastal ABL and variable wave state; and

• the role of remote mesoscale spatial inhomogeneities in controlling atmosphere–ocean dynamics in a coastal environment.

AIR QUALITY

Another important application of coastal meteorology is the prediction of pollutant dispersal. Our study covered issues relevant to the coastal environment. The highly variable winds near the coast may sweep pollutants out to sea on a land breeze but then bring them back with the sea breeze. More accurate estimates of the vertical motion fields associated with these wind systems are critical for determining the layers at which the pollutant will ultimately reside (and the horizontal direction in which it will move).

Further progress here would be helped by

• comprehensive tracer studies conducted at increasingly more complicated coastal sites (allowing for evaluation, validation, and eventual widespread use of improved dispersion models), and

• improved coordination between air pollution and boundary-layer field observation programs conducted on both sides of the littoral.

CAPABILITIES AND OPPORTUNITIES

Observations

The present observational network of routine in situ data is not adequate for most applications. The coastal rawinsondes, especially over the West Coast, are sparse. The buoy network is also sparse and only measures conditions near the surface. There are transient ship reports that supplement the buoy reports.

The National Oceanic and Atmospheric Administration's observational-equipment modernization will offer some improvements and some degradation. The first network Next Generation Weather Radar (NEXRAD) will provide an increase in over-water coverage: Doppler winds out to 150 km, reflectivity out to 400 km. Returns from the moving sea surface may possibly be interpreted to measure surface winds. No new rawinsondes are planned, and some coastal sondes may be moved inland. But efforts continue to use passive and active satellite techniques to infer the atmospheric and sea state. And surface-based remote sensors can give highly detailed spatial and temporal detail in the boundary layer.

Models

The emergence of high-performance workstations having substantial fractions of the calculation-speed performance and superior throughput of present-day mainframe supercomputers will allow researchers to run regional models with high resolution and to conduct numerous sensitivity studies.

Human Resources

It is the experience of the panel members that few universities have courses in the meteorology of coastal zones. Related areas of meteorological instrumentation and observational techniques are also underrepresented.

To improve our capabilities and opportunities, the Panel on Coastal Meteorology recommends:

• the use of recently developed remote sensors to obtain detailed, four-dimensional data sets to describe coastal regions and the upgrade of buoy and surface station networks to obtain quality, long-duration data sets;

• the on-site use of affordable, high-performance work stations that can provide decentralized computations during study of local phenomena, be used to determine the sensitivity of coastal processes to various influences, and be used to study techniques for assimilating data into real-time forecasts; and

• the increased use of conferences, short courses, and university training programs to encourage more scientists to explore the meteorology of the coastal zone.

3

Modeling Transport Processes in the Coastal Ocean

Alan F. Blumberg
Hydroqual, Incorporated
Richard P. Signell, Harry L. Jenter
U.S. Geological Survey

INTRODUCTION

As population growth and industrial development continue along the coastal zones, near-shore waters over the continental shelf are being subjected to increasing environmental stresses from numerous sources. Discharges of municipal and industrial wastes, agricultural runoff, combined sewer overflows, and waste spills of potentially toxic substances from coastal commerce contribute collectively to a host of water-quality problems. The eventual impact of these discharges is the result of complex interactions among the pollutant inputs from all sources; the various chemical forms of the constituents present in the water column and their associated chemical reactions with each other; and the complex marine food chains, which can exchange nutrients and other chemicals between the water column and the underlying sediment. To understand all of these interactions, it is necessary to define the hydrodynamic transport processes governing the movement and mixing of the constituents as they are forced by various hydrographical (runoff, estuarine circulation), meteorological (surface wind, heat fluxes), open ocean (large scale ocean circulation), astronomical (tides), and internal (density gradients) mechanisms.

The circulation occurring over the continental shelf typically exhibits considerable temporal and spatial variability. It is characterized by relatively large-scale alongshore current systems, which have a variety of interannual, seasonal, and daily variations. The variability in circulation from place to place is evident in satellite sea-surface temperature images that show patchy upwelling zones, filaments of cold water extending offshore, rotating eddies, and other large-scale circulation features. Processes responsible for the circulation are wind-driven currents and mixing, which are often the dominant processes over the shelf; buoyancy effects, which lead to plume and frontal formations; shelf-open ocean interactions, where meandering offshore currents and mesoscale eddies can entrain much water from the shelf; and tidal resonances, which can produce large tidal currents and intense levels of vertical mixing.

Observational programs have been the cornerstone of our conceptual and theoretical understanding of currents and water properties in coastal regions. Our knowledge has increased because of the introduction of moored, hydrographic, Lagrangian, and satellite-based observations. However, to permit a consistent view of the circulation, all of these types of observations are needed simultaneously. Consider, for example, that in many coastal regions the length scales of the hydrodynamical processes are characterized by an internal deformation radius (Rossby radius) of 5 km to 15 km and by topographic variations ranging from 1 km to 10 km. Motions and water properties measured at stations separated by distances much greater than the length of these scales will, in general, tend to be only loosely related to one another. The sampling networks of hydrographic surveys and current meter moorings must be chosen within the Rossby radius. For coastal domains of small extent, this is possible; however, for large regions it is not always feasible. There are also a host of issues associated with obtaining observations with the proper time scales. Satellites provide excellent spatial views that are, unfortunately, only *snapshots* in time. Current meters are better at addressing temporal variability, since they employ sampling frequencies that are typically 30 minutes or less; however, their spatial coverage is limited. It is apparent, then, that observational programs are rarely sufficiently dense (with respect to frequency of observation) in either space or time to provide an adequate description of the water mass and velocity fields of an evolving, three-dimensional piece of coastal ocean.

In recent years, coastal-ocean circulation models have come to be depended upon, when properly tested and verified, to synthesize information from measurements and provide a framework for investigating the basic processes of a region. As such, coastal-ocean models play a critical role in determining how nutrients, sediment, contaminants, and other water-borne materials are transported.

The purpose of this paper is to provide an overview of the processes governing the transport of water, dissolved substances, and particles in the context of the design and development of a three-dimensional circulation model of Massachusetts Bay. The bay is ideally suited for the purposes of this paper because its circulation is a complicated function of winds, tides, and river inflows, and because its environmental problems, caused by the introduction of wastes over many years, are typical of those found off many large metropolitan areas. Before the presentation of the Massachusetts Bay case study, a short discussion of numerical models themselves will be provided, briefly reviewing where coastal-ocean circulation modeling is today and describing the physically comprehensive circulation model developed by Blumberg and Mellor (1980) which will be used to elucidate the various transport processes in Massachusetts Bay. In the final section, some thoughts on the important issues of coastal-ocean modeling that need to be addressed in the future will be put forth.

COASTAL-CIRCULATION MODELING

Significant progress has been made in the development of limited-area coastal-circulation models. The state of the science has progressed to the point where programs to develop and validate a predictive system for the U.S. coastal ocean are being proposed (Joint Oceanographic, Inc., 1990). There are now models being used routinely in the Great Lakes to determine water levels, currents, and temperatures for periods going back 30 years (Bedford and Schwab, 1990). Much of the increase in modeling activities is due to the availability of low cost, supercomputer resources and the continued development of reliable numerical codes. Too many models exist to provide a comprehensive survey here. The interested reader is referred to the monographs by Heaps (1986) and Nihoul and Jamart (1987) and to the review articles by Wang et al. (1990) and Blumberg and Oey (1985) for details concerning the status of coastal ocean circulation modeling.

The coastal-ocean-circulation model developed by Blumberg and Mellor (1980), called ECOM3D, will be used as a framework for discussing the transport processes that operate in the coastal ocean. The model is three-dimensional and time dependent so that it can reproduce the complex oceanographic physics present over the shelf. Evolving water masses, baroclinic plumes, fronts, and eddies are accounted for by prognostic equations for the thermodynamic quantities, temperature, and salinity. Free surface elevation is also calculated prognostically so that tides and storm-surge events can be simulated. Through these prognostic equations and the use of the full, nonlinear form of the momentum equations, the processes relevant to a spectrum of nonlinear, stratified flows can be properly modeled. Coastal upwelling dynamics and the processes leading to stratified tidal rectification will be part of the simulated distributions. The vertical turbulent mixing processes are parameterized using the turbulent-closure submodel of Mellor and Yamada (1982). This submodel contains non-dimensional empirical constants that are fixed by reference to laboratory data and are independent of particular hydrodynamic model applications. ECOM3D also incorporates a σ-coordinate system such that the number of grid points in the vertical is independent of depth. This ensures that the dynamically important surface and bottom boundary layers across an entire sloping shelf can be adequately resolved. The last model feature to note is the use of a curvilinear coordinate system. This system greatly increases model efficiency in treating irregularly shaped coastlines and in meeting requirements for high resolution in specific local regions. A complete description of the governing equations and numerical techniques can be found in Blumberg and Mellor (1987). The model has been used in over 30 studies, which have appeared in the referred literature, and is being exercised in an operational forecasting mode for the Great Lakes and in Norwegian coastal waters.

A CASE STUDY: MASSACHUSETTS BAY

Massachusetts Bay and Cape Cod Bay combine to form a roughly 100 km x 50 km semi-enclosed basin with an average depth of 35 m located in the western Gulf of Maine (Figure 3.1). As in many coastal regions near major urban areas, the bays are used for a variety of purposes: commercial and recreational fishing, shipping, recreational boating, swimming, and as a repository for sewage effluent and dredged sediments. Currently, there is considerable controversy concerning the extension of the Boston sewage outfall pipe from the mouth of Boston Harbor to a site 9 miles offshore. The public living around the coast of Massachusetts Bay and Cape Cod Bay is concerned that Boston is improving its pollution problem at the expense of the bays and that swimming beaches, shellfish beds, fishing resources, and the endangered right whale population that feeds in the bays may be jeopardized. To address these concerns, ECOM3D is being used in conjunction with available observations to determine the fate and transport of contaminants, nutrients, and other water-borne materials in the bays, including effluent from the proposed outfall site. The region covered by the model encompasses

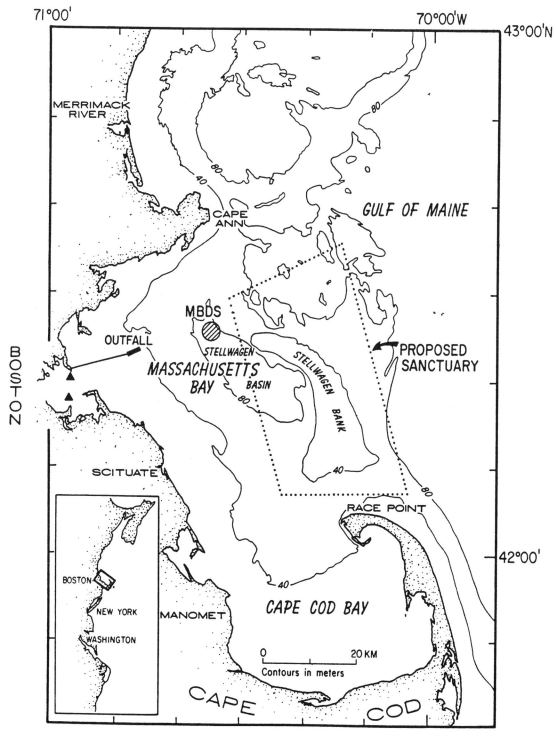

FIGURE 3.1 Bathymetric mapshowing Massachusetts and Cape Cod Bays, present sewage outfalls in Boston Harbor (solid triangles), location of new ocean outfall for treated Boston sewage in western Massachusetts Bay (average flow about 20 m³/s), approximate boundary of the proposed Stellwagen Bank Marine Sanctuary, and the Massachusetts Bay Disposal Site (MBDS).

all of Massachusetts Bay and Cape Cod Bay as well as Stellwagen Bank (Figure 3.2). It includes the Merrimack River and extends offshore to a depth of about 200 m. The resolution of the curvilinear grid system ranges from a minimum of 600 m near the proposed outfall site to a maximum of 6000 m near the open-ocean boundary. Vertical resolution is accomplished by using 10 a-levels in the water column. Data for model calibration and verification was obtained during an intensive field program over the period from 1990 to 1991 (Geyer et al., 1992).

Tidal Currents

The most predictable, and often the strongest, currents in the bays are produced by the barotropic tides, which have an average range of 2.6 m. Tides are introduced into the model by forcing the open offshore boundaries with sea surface elevation data from a well-calibrated, lower-resolution Gulf of Maine model (Naimie and Lynch, 1991). Comparison of modeled currents with moored-current observations from the winter (when the best analysis of the pure tidal signal can be made) reveals that both tidal excursions and orientations of tidal ellipses are reproduced well (Figure 3.3). Tidal excursions range from more that 12 km (off Provincetown and in Boston Harbor) to less than 2 km (in the deep central Massachusetts Bay). The tidal excursion at the proposed outfall is 2 km.

While the moored observations of tidal currents give some indication of the spatial variation, the model, verified at the moorings, can fill in the spatial structure. It indicates regions of strong tidal gradients where tidal mixing fronts may form as well as features unresolved by the observations (Figure 3.4). The locations of the strong gradients on Stellwagen Bank and west of Provincetown are near regions frequented by the endangered North Atlantic right whale (Hamilton and Mayo, 1988). Therefore, resolution of the flow in these regions may be important in determining the effects of pollution introduced into the bays on the whale population.

Because of their strength, tides play an important role in vertical mixing processes, but since they are periodic at 12.4 hours (the M_2 constituent dominates here), they essentially displace material back and forth over the length of the tidal excursion with little net transport. The exceptions are when tidal currents act in conjunction with the bottom topography and coastline geometry to produce strong asymmetry between ebb and flood. The tides have a large effect on the flushing of Boston Harbor (Signell and Butman, 1992) and may be important locally at the tip of Cape Cod and along the western side of Stellwagen Bank, but they have little impact on transporting material over distances comparable to the size of the bays.

Subtidal Currents

Observations suggest that horizontal transport of material is accomplished by advection due to the mean flow through the bays and the dispersive effect of subtidal wind-driven and river-runoff events. The mean flow generally supports the historical conceptual picture of a counterclockwise circulation (Bigelow, 1927; Bumpus and Lauzier, 1965; Brooks, 1985) made up of southwesterly inflow south

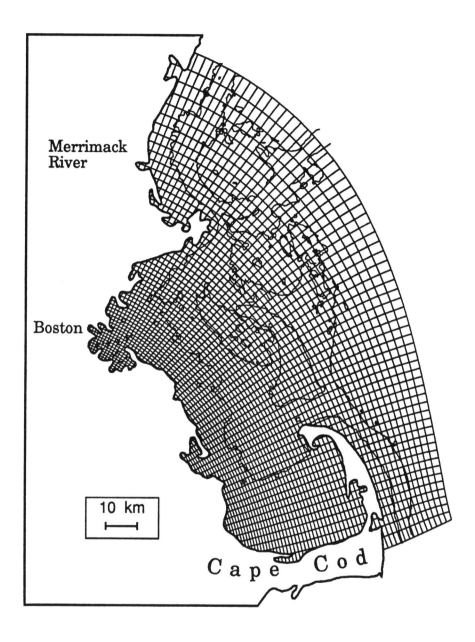

FIGURE 3.2 Model grid for the three-dimensional circulation model, ECOM3D, of Massachusetts Bay and Cape Cod Bay. The curvilinear orthogonal grid allows the mesh resolution to vary spatially, having a minimum grid spacing of 600 m and a maximum spacing of 6000 m. The grid spacing in the vicinity of the proposed outfall is roughly 1000 m. There are currently 10 vertical o-levels in the model, evenly spaced throughout the water column.

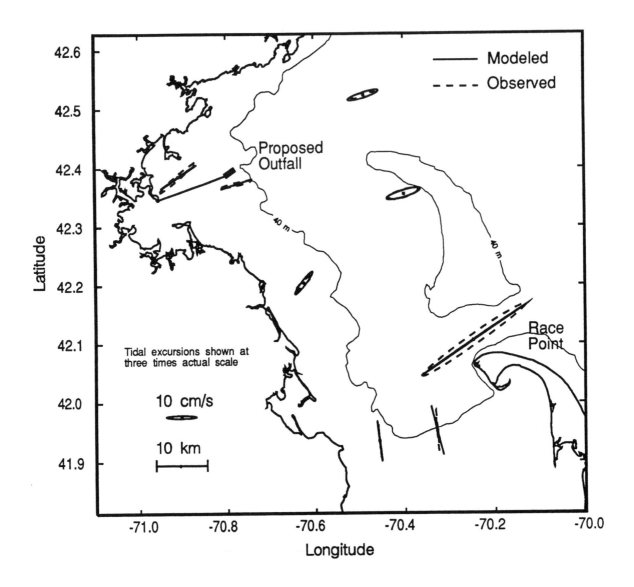

FIGURE 3.3   Comparison of modeled and observed surface M2 barotropic tidal currents in Massachusetts Bay. Shown are tidal ellipses, which indicate the observed velocities over the tidal cycle. They also represent the excursions water parcels would make if they moved with the tidal currents observed at the mooring. For clarity these tidal excursions are shown at three times actual scale. Tidal excursions are nearly 10 km off Race Point but only 2 km near the location of the proposed outfall.

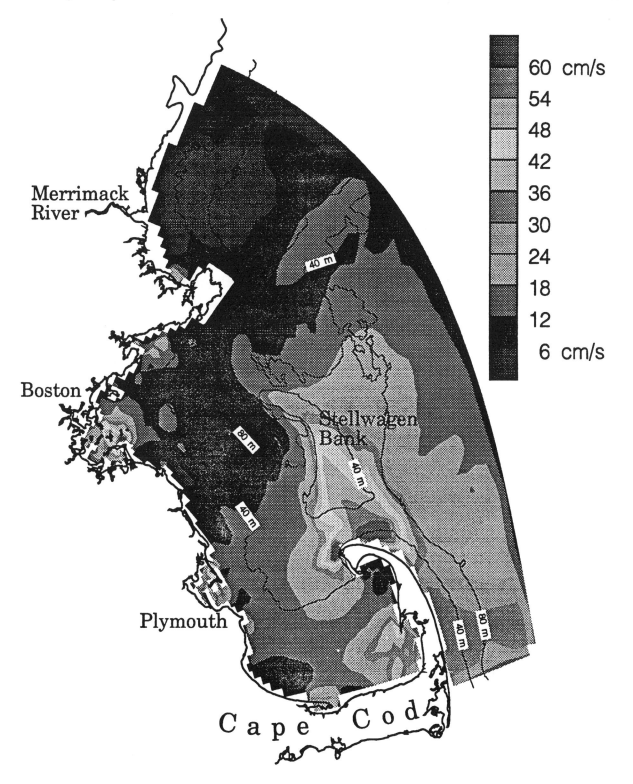

FIGURE 3.4 Maximum tidal-current speed at the water surface off the tip of Cape Cod and in the entrances of smaller embayments such as Boston and Plymouth Harbors.

of Cape Ann, southerly flow along the coast east of Scituate, and northeasterly outflow north of Race Point (Figure 3.5). This mean circulation pattern, however, is often altered by wind and runoff events, and except at the deep stations near Cape Ann and Race Point, the fluctuations are typically stronger than the mean. The proposed outfall site, in fact, is in a region of weak mean flow, west of the stronger residual current system. This means that material here is mixed and transported by random fluctuations of wind and runoff events rather than being swept away by a persistent current. Using process-oriented modeling that examines the bays' response to specified forcing conditions, the factors that are important for driving the mean flow and the low-frequency fluctuations can be determined.

Mean Flow

What drives the observed counterclockwise flow through the bays? One hypothesis is that it is simply an extension of the coastal current that exists in the Gulf of Maine (Bigelow, 1927; Bumpus and Lauzier, 1965; Vermersch et al., 1979). To test this hypothesis, the model is forced with a 3-cm offshore sea-surface slope from the coast to the 100-m isobath along the northern boundary. This slope produces a 10-cm/s coastal current north of Cape Ann that is comparable to observed coastal current speeds (Vermersch, 1979). The simulation reveals that much of the Gulf of Maine coastal current moves southward following the bathymetry along the eastern flank of Stellwagen Bank, largely bypassing the bays (Figure 3.6). The coastal current explains the observed mean flow southeast of Cape Ann and at Stellwagen Bank (stations U2, U3 and U6 in Figure 3.5), but the counterclockwise flow that the coastal current drives in the bays is much weaker than observed. Adding the mean wind stress of 1 dyne/cm^2 to the coastal current forcing dramatically changes the simulation of the mean flow. The mean wind is from the west and drives a realistic, simulated southeastward current along the coast from Boston to Cape Cod, which exits the bays at Race Point (Figure 3.7). Thus, remote forcing from the Gulf of Maine coastal current and the mean wind stress both play important roles in explaining the mean circulation in the bays.

Low-Frequency Fluctuations

The wind direction is one factor that determines the response of the bays: northwest or southeast winds are aligned with the long-axis of the bays and are therefore more efficient at driving circulation than southwest or northeast winds (Geyer et al., 1992). When the waters of the bays are unstratified, as in winter, northwest winds drive strong flow downwind at the coast, which piles up water in Cape Cod Bay, creating an along-bay pressure gradient (Figure 3.8). This pressure gradient drives return flow against the wind at depth (Figure 3.9). When the wind blows from the southwest during well-mixed conditions, the currents are substantially weaker (Figure 3.10).

Another factor that has a major impact on the wind response is the degree of stratification. Results from two years of measurements near the proposed outfall show that the surface current fluctuations

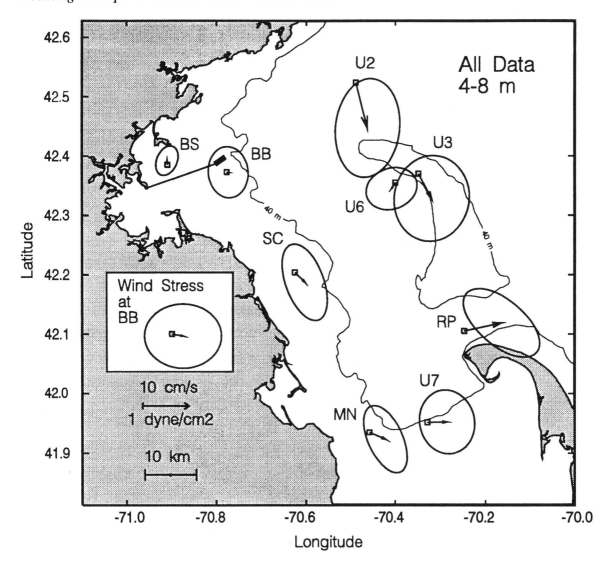

FIGURE 3.5 Map showing the mean flow (solid arrow) and the low-frequency variability (shown as ellipses centered around the tip of the mean flow) for all near-surface (4-8 m depth) current measurements made from December 1989 to September 1991. The daily averaged current originates at the station symbol (two-character names) and flows toward any location within the ellipse. The arrows and ellipses have been scaled to correspond to the distance a particle moving with that current would travel in one day.

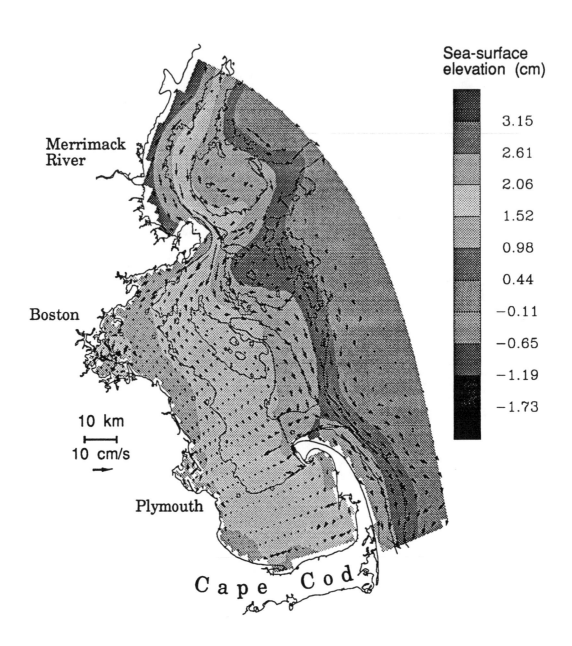

FIGURE 3.6 Modeled surface current and elevation response to the Gulf of Maine coastal current.

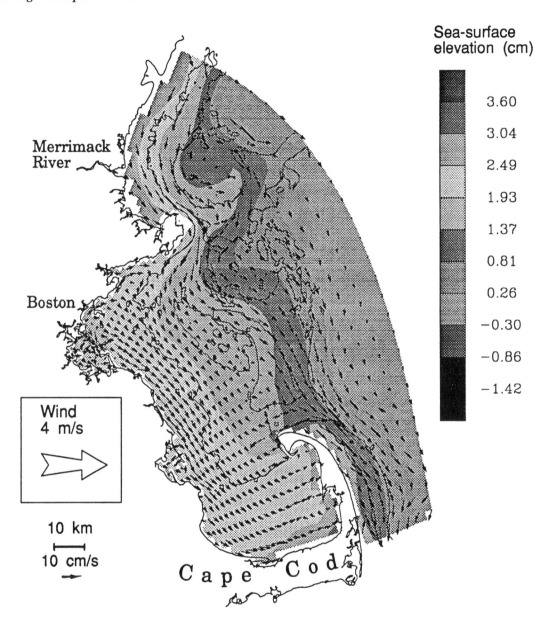

FIGURE 3.7 Modeled surface current and elevation response to the Gulf of Maine coastal current and the mean wind stress.

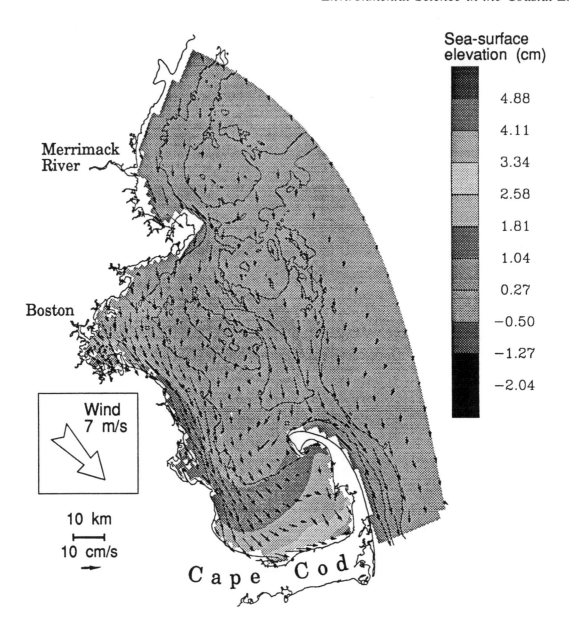

FIGURE 3.8 Modeled surface current and elevation response to a wind stress from the northwest of 1 dyne/cm² (about 7 m/s) in well-mixed conditions. This along-bay wind drives strong flow downwind in the shallow water near the coast. The convergence of surface water along the northern shore of Cape Cod indicates the presence of strong downwelling.

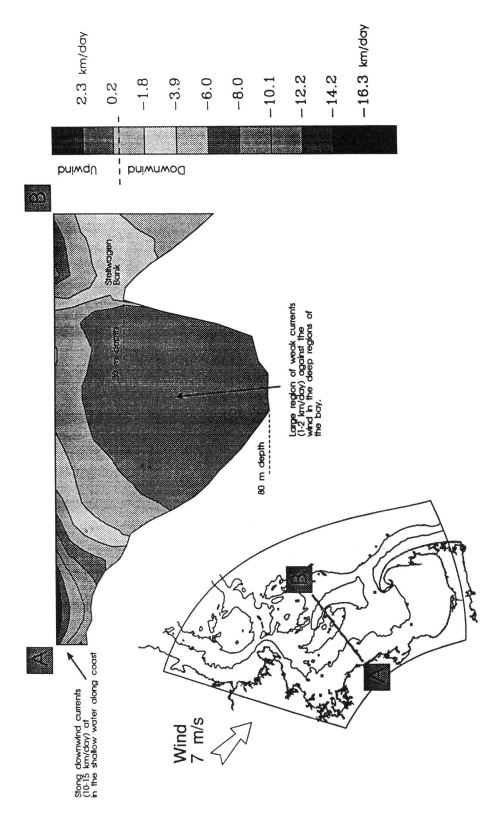

FIGURE 3.9 Vertical section of along-bay current response to wind stress from the northwest of 1 dyne/cm² (about 7 m/s) in well-mixed conditions. The along-bay wind drives downwind currents of 10-15 cm/s in the shallow water near the coast. The downwind current establishes an opposing pressure gradient, which drives a weak return flow (1-2 cm/s) at depth.

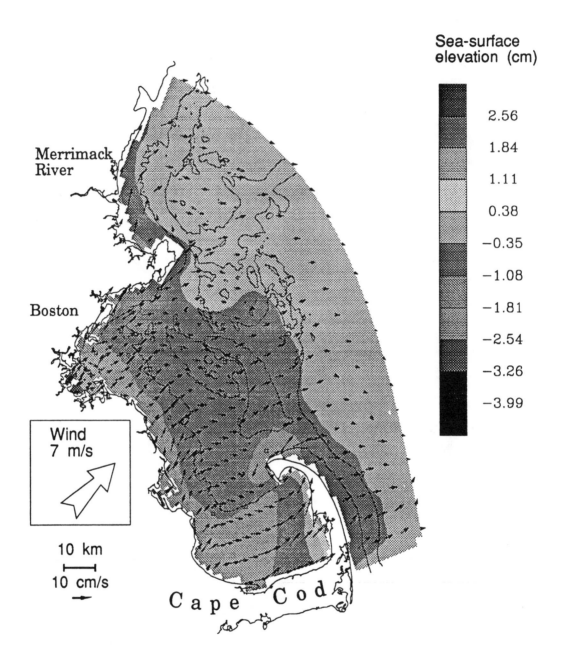

FIGURE 3.10 Modeled surface current and elevation response to a wind stress from the southwest of 1 dyne/cm^2 (about 7 m/s) in well-mixed conditions. The cross-bay wind drives downwind surface flow of nearly the same magnitude over most of the Massachusetts Bays. The currents near the coast are significantly weaker than those generated by the along-bay wind shown on Figure 3.8.

are strongest in summer, when the wind-stress fluctuations are weakest (Figure 3.11). One hypothesis regarding this is that the wind stress is more efficient at driving surface currents in summer, when strong stratification reduces the frictional resistance of the surface layer. Modeling the response of the bays to southwest wind under stratified conditions (Figure 3.12) reveals a much different picture than the unstratified case (Figure 3.10). The surface currents are often twice as large, and the circulation pattern is dramatically different. As the water moves offshore along the coast north of Boston, cold nutrient-rich water is upwelled, as evident in the model, as well as in remotely sensed images of sea-surface temperature (Figure 3.13). Being able to model this type of response is especially important for understanding the proposed outfall's impact, as the outfall plume may be trapped in the cold, light-limited, deeper waters until it is upwelled.

Although the Massachusetts bays do not have any large rivers that discharge directly into them, the Merrimack River just to the north of Cape Ann plays an important role in driving their circulation, especially in the spring (Butman, 1975). In the absence of wind, fresh water discharged from the river mouth forms a surface plume that turns to the right and follows the coast in the northern hemisphere. Thus the Merrimack and other Gulf of Maine rivers combine to generate a buoyancy-driven coastal current that flows southward off Cape Ann. Large river runoff events can drive currents with magnitudes of 20-40 cm/s in the bays, which are magnitudes comparable to those during strong wind events.

Wind and river effects often interact nonlinearly, generating extremely complex flow patterns, even for fairly simple forcing functions. As an example, *turning on* the Merrimack River under the influence of a mean wind from the west gives rise to lateral and vertical salinity gradients that act with the wind and local topography to yield small-scale lateral eddying structures in western Massachusetts Bay that change markedly with time and position in the water column (Figure 3.14). This illustrates the difficulty with isolating different forcing mechanisms from field data. Often, the mechanisms cannot be linearly superimposed. Numerical models frequently become the only viable mechanism for analyzing the transport of material when this is the case.

Outfall Plume Dynamics

Because of the complicated nature of the circulation in embayments such as the one considered here, it is difficult to make simple calculations based on observations regarding the transport of material suspended or dissolved in the water column. In particular, it is clear that realistically modeling the effluent plume from the proposed outfall requires a three-dimensional model that actively couples the density field with the circulation. Because the plume is buoyant and may enter a stratified system, the model must have the ability to allow the effluent to become trapped below the thermocline or to rise to the surface as determined by the ambient stratification.

a)

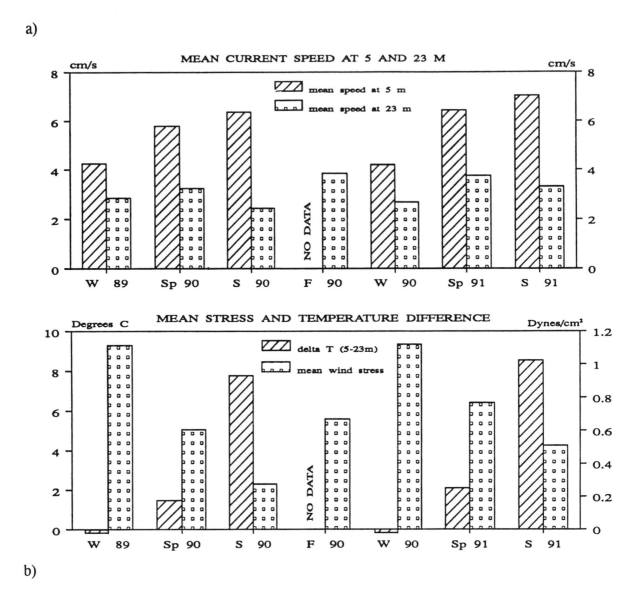

b)

FIGURE 3.11 a) Mean daily vector-averaged current speed at Station BB (Figure 3.5) at 5-m and 23-m water depth (winter, November through February; spring, March through May; summer, June through August; fall, September and October). b) Mean daily averaged wind stress amplitude at the Large Navigational Buoy and temperature difference between 5-m and 23-m water depth. These data suggests that stratification is more important than wind in determining the strength of the near-surace current at the new outfall site.

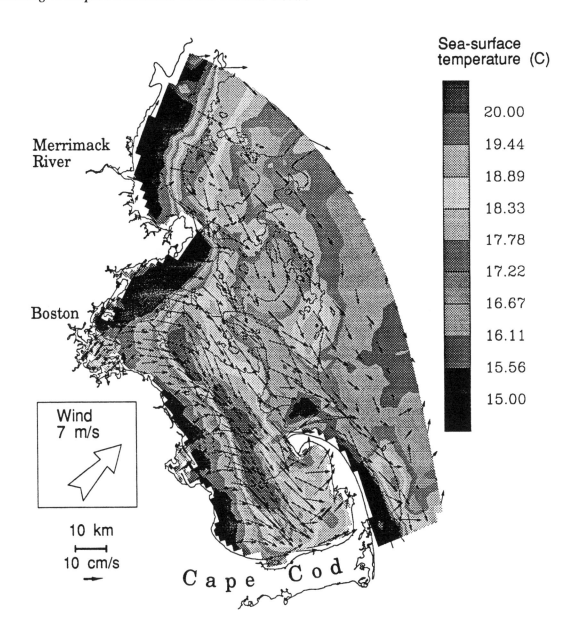

FIGURE 3.12 Modeled surface current and sea-surface temperature respond to a wind stress from the southwest of 1 dyne/cm^2 (about 7 m/s) in stratified conditions. From a homogeneous state, the model was run with M$_2$ tides and a surface heat flux of 100 W/m^2 for 12 days. The wind was imposed for the last two days of the simulations. The wind-induced upwelling brings cold, nutrient-rich water from depth to the surface and plays an important role in primary production.

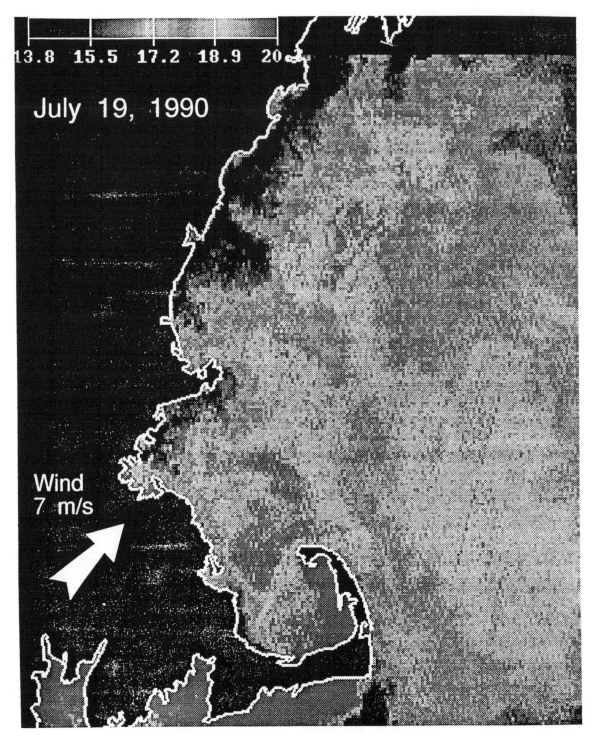

FIGURE 3.13 Observed sea-surface temperature during a typical summer upwelling event, at 0600 on July 19, 1990. On the preceding two days, the wind was relatively constant at a speed of about 7 m/s from the southwest. The observed surface-temperature patterns are quite similar to the modeled upwelling event, especially along the coast as shown in Figure 12.

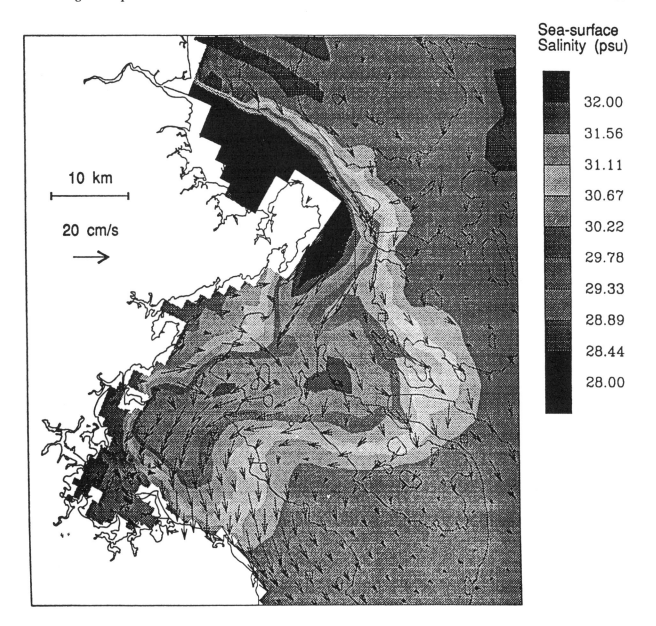

FIGURE 3.14 Modeled surface current and salinity response of western Massachusetts Bay to a runoff event from the Merrimack River. The Merrimack was *turned on* with a flow of 500 m³/s, a magnitude typical of a spring freshet (Geyer, 1992), and shown is the result after 8 days. The runoff event creates strong and complex currents in the vicinity of the proposed outfall site, which would be difficult to resolve with moored instruments.

This is critical, as the surfaces above and below the thermocline are often moving in opposite directions. Consider the movement of a plume produced through the discharge of 15 m³/s of fresh water into a well-mixed Massachusetts Bay with a salinity of 32 psu. A steady wind stress of I dyne/cm² (about 7 m/s) yields the flow patterns shown in Figures 3.8 and 3.9 and also causes a complex plume structure (Figure 3.15). As the plume rises to the surface, it is advected northwestward by the bottom currents. However, as the plume approaches the surface, it is advected with the oppositely directed surface currents to the southeast.

FUTURE DIRECTIONS AND CONCLUDING REMARKS

Our knowledge of the processes affecting the movement and mixing of water masses, various chemical constituents, and particles and our ability to model them numerically has expanded considerably in the past five years. Hydrodynamic models of the coastal ocean have become indispensable aids in decision making relating to wasteload allocations from point sources of pollution and to the design and licensing of offshore structures. The models typically are at their best in predicting phenomena related to the tides and to the response of the waters to local forcing conditions. However, strict attention must be given to the adequacy of the model grid-spacing, because experience has shown that when the model resolution is commensurate with the physical process of a region, the model simulations agree best with the observations.

It must be mentioned that the most critical factor limiting the development of truly predictive models is an understanding of the complex interaction between the coastal waters and those of the offshore ocean. Typically, coastal ocean models cover a limited region along the coast with their offshore extent ending at the continental shelf break. At this edge, it is necessary to introduce a boundary condition that properly parameterizes the influences of the ocean exterior to the coastal region being modeled. The use of both high-quality data sets and specially designed numerical model experiments are needed to determine the proper feedbacks between the two regions.

Advances in computer power and speed have recently made it feasible to construct and apply time-varying, three-dimensional hydrodynamic models. This type of modeling, while feasible, is computationally intensive and puts severe constraints on the resources available for engineering analyses. What is needed are models that include all the important physical, chemical, and biological processes yet can be used in a time effective manner without significantly depleting the available computer resources. One approach to this problem is through the development of a hydrodynamic and water-quality-model interfacing methodology to produce time averaged, residual transports of coarse spatial resolution from an intratidal hydrodynamic model of high spatial resolution. These time and space averaged mass transport quantities should be sufficient to drive appropriate segmented water-quality models without loss of accuracy.

Finally, there is a need to make better use of available observations. Models require data to establish interior and boundary conditions, to update boundary fields, to validate the model physics, and to verify the simulations. Current, temperature, and salinity data are often insufficient for an unambiguous model calibration/validation and one must look at how well the water-quality constituents

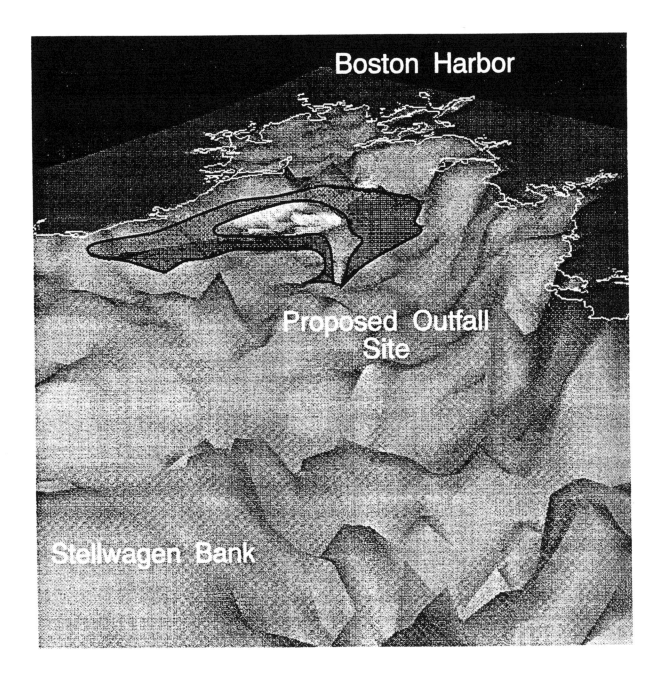

FIGURE 3.15 Three-dimensional view of the plume caused by a discharge of 15 m^3/s of fresh water at the proposed outfall site under well-mixed conditions. A wind stress of 1 dyne/cm^2 (about 7 m/s) was applied from the west, and shown are dilution isosurfaces of 250 and 500 after 9 days of simulation. As the plume rises to the surface, it is pulled toward Boston by the bottom currents, but when it reaches the surface it moves with the surface currents and is transported along the coast toward Cape Cod.

are being modeled to derive a sense of the validity of the modeled transport processes. One needs to blend the results from both circulation and water quality models with the available data to provide for the best estimates of how water and materials are transported throughout a coastal system. The data assimilation, that is, the process of this blending, is undoubtedly the most powerful tool presently available for extracting information and insight from the sparse coastal-ocean data sets and the imperfect model results.

ACKNOWLEDGEMENTS

This work has been funded by the Massachusetts Water Resources Authority through contracts with both HydroQual (Marine Technical Environmental Services Contract #37 to Normandeau Associates) and the U.S. Geological Survey. Additional support has been provided by the Massachusetts Environmental Trust, the U.S. Environmental Protection Agency's Massachusetts Bays Program, and the U.S. Geological Survey. The authors would like to thank Robert Beardsley for his role in motivating this paper. This paper also appears in the Journal of Marine Environmental Engineering, 1:31-52, 1993, published by Gordon and Breach, Reading, England.

REFERENCES

Bedford, K., and D. Schwab. 1990. Preparation of Real-Time Great Lakes Forecasts. CRAYCHANNELS, Cray., Res., Inc., p. 14-17.

Bigelow, H. B. 1927. Physical oceanography of the Gulf of Maine. Bull. U. S. Bur. Fish. 40(511):1027.

Blumberg, A. F., and G. L. Mellor. 1980. A coastal ocean numerical model. Pp. 203-214 in: Mathematical Modelling of Estuarine Physics, Proceedings of the International Symposium, Hamburg, 24--26 August 1987. Berlin: Springer-Verlag.

Blumberg, A. F., and G. L. Mellor. 1987. A description of a three-dimensional coastal ocean circulation model. Pp. 1-16 in: Three-Dimensional Coastal Ocean Models, Coastal and Estuarine Sciences, 4. Washington, D.C.: American Geophysical Union.

Blumberg, A. F., and L. Y. Oey. 1985. Modeling circulation and mixing in estuaries and coastal oceans. Pp. 525-547 in Advances in Geophysics, 28A.

Brooks, D. 1985. Vernal circulation in the Gulf of Maine. Journal of Geophysical Research.

Bumpus, D., and M. Lauzier. 1965. Circulation on the Continental Shelf of the East Coast of Eastern North America between Newfoundland and Florida. Washington, D.C.: American Geographical Society Serial Atlas of the Marine Environment, Folio 7.

Butman, B. 1975. On the Dynamics of Shallow Water Currents in Massachusetts Bay and the New England Continental Shelf Ph.D thesis, Massachusetts Institute of Technology and Woods Hole Oceanographic Institution.

Butman, B., M. Bothner, J. Hathaway, H. Jenter, H. Knebel, F. Manheim, and R. Signell. 1992. Contaminant Transport and Accumulation in Massachusetts Bay and Boston Harbor: A Summary of U.S. Geological Survey Studies. USGS Open-File Report 92-202.

Joint Oceanographic Inc. 1990. Coastal Ocean Prediction Systems Program. Available from Joint Oceanographic Inc., Suite 800, 1755 Massachusetts Avenue, NW, Washington DC 20036-2102.

Geyer, W., G. Gardner, W. Brown, J. Irish, B. Butman, T. Loder, and R. Signell. 1992. Physical Oceanographic Investigation of Massachusetts and Cape Cod Bays, Technical Report WHOI-92-x (in press), Woods Hole, Massachusetts: Woods Hole Oceanographic Institution.

Hamilton, P., and C. Mayo. 1988. Population characteristics of Right Whales (Eubal aena glacialis) observed in Cape Cod and Massachusetts Bays, 1978-1986. Rep. Int. Whal. Commn (Special Issue 12):203-208.

Heaps, N. S. 1986. Three-Dimensional Coastal Ocean Models, Coastal and Estuarine Sciences, 4, Washington, D.C.:American Geophysical Union.

Mellor, G. L., and T. Yamada. 1982. Development of a turbulence closure model for geophysical fluid problems. Rev. Geophys. Space Phys., 20:851-875.

Nihoul, J. C. J., and B. J. Jamart. 1987. Three-Dimensional models of Marine and Estuarine Dynamics. Amsterdam: Elsevier Scientific Publishing Co.

Naimie, C., and D. Lynch. 1991. Benchmark 3-D M2 and M2 Residual Tides for Georges Bank and the Gulf of Maine, Technical Report NML-91-2. Numerical Methods Laboratory, Thayer School of Engineering, Dartmouth College. Hanover, New Hampshire: Dartmouth College.

Signell, R. P., and B. Butman. 1992. Modeling tidal exchange and dispersion in Boston Harbor, accepted to Journal of Geophysical Research.

Wang, J. D., A. F. Blumberg, H. L. Butler, and P. Hamilton. 1990. Transport prediction in partially stratified tidal water. Journal of Hydraulic Engineering, 116:380-396.

Vermersch, J., R. Beardsley, and W. Brown. 1979. Winter circulation in the western Gulf of Maine: Part 2. Current and pressure observations. Journal of Phys. Oceanography 9:786-784.

4

Coastal Geomorphology

Stephen P. Leatherman, A. Todd Davison, Robert J. Nicholls
University of Maryland
College Park, Maryland

INTRODUCTION

Coastal geomorphology, by definition, is the study of the morphological development and evolution of the coast as it acts under the influence of winds, waves, currents, and sea-level changes. This study of physical processes and responses in the coastal zone is often applied in nature, but it also involves basic research to provide the fundamental understanding necessary to address the pertinent questions.

A principal coastal concern today and in the foreseeable future is beach erosion. It is estimated that 70 percent of the world's sandy shorelines are eroding (Bird, 1985). In the United States the percentage may approach 90 percent (Leatherman, 1988). This worldwide extent of erosion suggests that eustatic sea-level rise is an important underlying factor, although many other processes contribute to the problem. In many low-lying coastal areas, human impacts, such as the maintenance of tidal inlets and subsidence induced by groundwater and hydrocarbon withdrawals, have also made a substantial contribution to the erosion problem (National Research Council, 1990). At the same time, coastal populations are burgeoning, and this trend seems set to continue (Culliton et al., 1990). This raises the fundamental question -- what is the best response to the problem of shoreline recession?

Faced with progressive shoreline retreat and the inevitable loss of protective and recreational beaches, coastal communities have three basic alternatives: (1) retreat (relocate buildings and other infrastructure in a landward direction), (2) accommodate (e.g., raise buildings to the projected higher flood levels), or (3) protect (build hard structures or use beach nourishment methods). In areas of dense population and highly developed infrastructure, protection is the preferred alternative. Hard structures are costly and inflexible and often have environmentally and aesthetically undesirable effects such as loss of the recreational beach. Thus, beach nourishment has become the coastal management *tool of choice* over the last several decades (Leatherman, 1991).

To date, it is estimated that over 640 km of U.S. coastline have been nourished, largely through public funding, at a total cost of about $8 billion (Dixon and Pilkey, 1989). The use of beach nourishment as a coastal management tool will probably continue its significant growth over the next few decades. The contemplated economic commitments to this management alternative by federal, state, and local governments is unprecedented. For instance, in northern New Jersey a Congressionally authorized nourishment project proposes to reinstate 19 km of beach at a cost of approximately $200 million with projected maintenance costs over 50 years of about $300 million (Bocamazo, 1991). Similarly, the total cost of the recently (1991) completed Ocean City, Maryland, nourishment project, including renourishment every four years for 50 years, is estimated at $342 million (Kelly, 1991).

Predictability of the performance of beach nourishment is still poor in spite of its increasing use. This lack of understanding exists because: (1) predictive models of beach behavior in response to varying hydrodynamic forces are still relatively crude tools for engineering purposes and (2) most completed projects did not include adequate post-emplacement monitoring to allow for objective project assessment and necessary adjustment of designs (Davison et al., 1992). Therefore, each beach fill remains, in part, an educated experiment. Although many believe that there is sufficient understanding and inherent flexibility within the procedure to produce practical and successful designs (Delft Hydraulics, 1987), this confidence is not universally accepted.

During the 1980s, because of the actual or perceived failure of numerous projects, beach nourishment began receiving heavy criticism as an ill-advised use of taxpayers' money (e.g., Gilbert, 1986). During this time, several researchers (e.g., Leonard et al., 1989) began to contradict the traditional coastal engineering methods used to design and evaluate such projects. Such criticisms are not isolated, and many coastal environmental groups advocate planned retreat as the only true solution to coastal erosion.

The conclusions of Leonard et al. (1989) have been challenged by many in the scientific and engineering communities (e.g., Strine and Dalrymple, 1989; Houston, 1991a). Nonetheless, contentions from the Pilkey camp have focused attention on the lack of high-quality monitoring of U.S. beach nourishment projects and acted as a catalyst for renewed research efforts. This controversy places beach nourishment in the forefront of public policy decisions in the coastal zone. The basic aim of beach nourishment is to advance the shoreline a given distance and hence realize all the consequent benefits such as increased storm protection. Accurate designs are essential for predicting beach-fill longevity and maintenance requirements, which necessitates quantitative understanding of the transport processes. Other pertinent questions involve the volume and grain size of sand required to attain a specific increase in subaerial beach width. Also, what is the lifetime and thus the renourishment frequency of a particular beach?

A major problem is that many of the present design concepts remain relatively untested against actual field performance. A Delft Hydraulics (1987) report summarizes our present understanding: "an exact forecast of the behavior of the beach fill is not possible, not even in the case where a large number of data of the relevant areas is available". At the present stage of technology, beach nourishment is more art than science (Egense and Sonu, 1987). The behavior of nourished and natural beaches is subject to the same uncertainties, and Wiegel (1987) argues that our present inadequate quantitative knowledge of natural beach processes handicaps dependable estimates on how well nourished beaches will perform. From a fundamental perspective, future shoreline evolution will always be stochastic, even with complete understanding of all the processes, because the underlying driving forces (e.g., waves, storms) are themselves stochastic (National Research Council, 1990). Thus, probabilistic predictions of nourishment performance must be the goal.

Evaluation of beach nourishment projects requires knowledge of cross-shore sand transport limits as well as delineation of the profile of equilibrium; these are fundamental concepts in coastal geomorphology (Leatherman, 1991). It is not often appreciated that most of the active beach profile is submerged. The entire profile must be moved seaward for nourishment to be successful. Thus, the seaward limit of the active beach profile for the purposes of beach nourishment is a problematic but very important determination (Bruun, 1986). Early nourishment projects did not consider the offshore profile (Jarrett, 1987), or if they did, utilized unrealistic slopes, which caused excessive losses of the subaerial beach (Hansen and Lillycrop, 1988). Hallermeier (1981) developed a wave-based profile zonation, including the depth definition (d_l), which he proposed as the seaward limit for beach-fill design. Limited field observations support this recommendation (Houston 1991b). The equilibrium profile concept can also be applied to beach nourishment design (Dean, 1983; 1991). But clearly, more field data are required, and routine post-project monitoring should include measuring the entire active profile to the depth of closure, which is generally less than 10 meters of water depth on U.S. coasts. Such basic data as time-series surveys of beach profiles are difficult or impossible to obtain for most of the 155 nourished beaches considered by Pilkey (1988). Therefore, the effectiveness of beach nourishment projects, particularly actual versus predicted performance, is debatable, and substandard sources, such as the local media, have been used to declare project success or failure.

Another problem involved in assessing the performance of beach nourishment is the widespread lack of post-project monitoring by independent, objective parties. Conflicting statements concerning the success or otherwise of beach nourishment are common in the literature (Davison et al., 1992). It is clear that project performance can only be objectively assessed if high quality monitoring data are available and considered using commonly agreed upon criteria of success and failure (Stauble and Hoel, 1986).

There is also frequently a lack of commitment or inability of project sponsors to properly maintain nourished beaches. This raises important questions about the accreditation of beach nourishment projects, particularly when such projects are being used as a means to potentially lower 100-year flood levels and hence to reduce the cost of federal flood insurance. Also, many states now petition the Federal Emergency Management Agency for funds to restore their eroded beaches after Presidential disaster declarations. Clearly there need to be established criteria for design, maintenance, and financial commitment for the accreditation of beach nourishment projects (Davison, 1992; Davison et al., 1992).

The increasingly developed character of the nation's coastline will undoubtedly lead to increasing demand for beach nourishment. It is hoped that this will be undertaken within the context of sensible management plans. In addition to population and development pressure, accelerated sea-level rise will also increase the demand for beach nourishment (Weggel, 1986; Leatherman and Gaunt, 1989; Stive et al., 1991). This raises a number of new questions, particularly regarding the seaward limit of the beach profile over long time scales and the long-term availability of sufficient sand. These fundamental concepts in coastal geomorphology will undoubtedly receive considerable attention in the coming decades.

REFERENCES

Bird, E. C. F. 1985. Coastline Changes--A Global Review. Chichester, England: John Wiley-Interscience, 219 pp.

Bocamazo, L. 1991. Sea Bright to Manasquan, New Jersey Beach erosion control projects. Shore and Beach 59(3):37-42.

Bruun, P. 1986. Sediment balances (land and sea) with special reference to the Icelandic south coast from Torlakshofen to Dyrholarey. River nourishment of shores--Practical analogies on artificial nourishment. Coastal Engineering 10:193-210.

Culliton, T. J., M. A. Warren., T. R. Goodspeed, D. G. Remer, C. M. Blackwell, and J.J. McDonough, III. 1990. 50 Years of Population Change Along the Nation's Coasts 1960-2010. National Ocean Service. Rockville, Maryland: National Oceanic and Administration.

Davison, A. T. 1992. The National Flood Insurance, Mitigation, and Erosion Management Act of 1991: Background and overview. In Proceedings of the National Conference on Beach Preservation Technology '92. Tallahassee, Florida: Florida Shore and Beach Preservation Association.

Davison, A. T., R. J. Nicholls, and S. P. Leatherman. 1992. Beach nourishment as a coastal management tool. Journal of Coastal Research 8:984-1022.

Dean, R. G. 1983. Principles of Beach Nourishment. In P.D. Komar, ed., CRC Handbook of Coastal Processes and Erosion. Boca Raton, Florida: CRC Press Inc.

Dean, R. G. 1991. Equilibrium beach profiles: Characteristics and applications. Journal of Coastal Research 7:53-84.

Delft Hydraulics. 1987. Manual on Artificial Beach Nourishment. Centre for Civil Engineering Research, Codes and Specifications, Rijkswaterstaat, Report 130.

Dixon, K. and O. H. Pilkey. 1989. Beach Replenishment on the U.S. Coast of the Gulf of Mexico. Pp. 2007-2020 in American Society of Civil Engineers: Proceedings of Coastal Zone '89 Conference, New York, New York.

Egense, A. K. and C. J. Sonu. 1987. Assessment of beach nourishment methodologies. Pp. 4421-4433 in Proceedings Coastal Zone '87. American Society of Civil Engineers, New York.

Gilbert, S. 1986. America washing away. Science Digest 94:28-35, 75, 78.

Hallermeier, R. J. 1981. A profile zonation for seasonal sand beaches from wave climate. Coastal Engineering 4:253-277.

Hansen, M. and W. J. Lillycrop. 1988. Evaluation of closure depth and its role in estimating beach fill volumes. Pp. 107-114 in Proceedings Beach Preservation Technology '88, Florida Shore and Beach Preservation Association, Florida.

Houston, J. R. 1991a. Rejoinder To: Discussion of Pilkey and Leonard (1990) [Journal of Coastal Research, 6, 1023 et seq.] and Houston (1990) [Journal of Coastal Research, 6, 1047 et seq.]. Journal of Coastal Research 7: 565-577.

Houston, J. R. 1991b. Ocean City, Maryland, beachfill performance. Annual Meeting of the American Shore and Beach Preservation Association, October 18, 1990 (Atlantic City, New Jersey). Shore and Beach 59:15-24.

Jarrett, J. T. 1987. Beach nourishment -- A Corps perspective, U.S. Army Corps of Engineers, Coastal Engineering Research Board 48th Meeting, Savannah, Georgia. District, Pp. 1-3.

Kelly, Jam. Gen. P. 1991. Keynote Address: America Shore and Beach Preservation Association Annual Meeting, October 17, 1990, Atlantic City, New Jersey Shore and Beach, 59(3):3-6.

Leatherman, S. P. 1988. Beach response strategies to accelerated sea-level rise. Pp. 353-358 in proceedings 2nd North American Conference on Preparing for Climate Change. Washington, D.C.: The Climate Institute.

Leatherman, S. P. 1991. Coast and Beaches. Pp. 183-200 in Kiersch, G.A., ed., The Heritage of Engineering Geology; The First Hundred Years. Centennial Special Vol. 3, Boulder, Colorado: Geological Society of America.

Leatherman, S. P. and C. H. Gaunt. 1989. National assessment of beach nourishment requirements associated with accelerated sea level rise. American Society of Civil Engineers: Proceedings of Coastal Zone '89 Conference, New York, NY, pp. 1978-1993.

Leonard, L. A., T. D. Clayton, K. L. Dixon, and O. H. Pilkey. 1989. U.S. beach replenishment experience: A comparison of beach replenishment on the U.S. Atlantic, Pacific, and Gulf of Mexico coasts. Pp. 1994-2006 proceedings Coastal Zone '89, American Society of Civil Engineers, New York.

National Research Council. 1990. Managing Coastal Erosion. Marine Board, National Research Council, Washington, D.C.: National Academy Press.

Pilkey, O. H. 1988. A "thumbnail method" for beach communities: Estimation of long-term beach replenishment requirements. Shore and Beach 56:23-31.

Stauble, D. K. and J. Hoel. 1986. Physical and Biological Guidelines for Beach Restoration Projects: Part II-Physician engineering Guidelines. Report Number 77. Gainesville, Florida Sea Grant College 100 pp.

Stive, M. J. F., R. J. Nicholls, and H. J. DeVrind. 1991. Sea-level rise and shore nourishment: a discussion. Coastal Engineering 16:147-163.

Strine, M. A., and R. A. Dalrymple. 1989. Beach Fill at Fenwick Island, Delaware. Center for Applied Coastal Research Department of Civil Engineering University of Delaware. Report 89-01. 61 pp.

Weggel, J. R. 1986. Economics of beach nourishment under scenario of rising sea level. Journal of Waterway, Port, Coastal and Ocean Engineering 112:418-427.

Wiegel, R. L. 1987. Trends in coastal engineering management. Shore and Beach, 55(1):2-11.

5

Rivers and Estuaries: A Hudson Perspective

Richard F. Bopp, Daniel C. Walsh
Rensselaer Polytechnic Institute
Troy, New York

INTRODUCTION

The perspective of this paper and the specific examples used for illustration reflect the fact that the principal author has spent much of the past fifteen years involved in geochemical research on the Hudson River and contiguous waters. The general observations should be broadly applicable to rivers and estuaries and provide a basis for further discussion.

The initial announcement of the Coastal Zone Retreat proposed a goal of "approaching broad resource and environmental issues somewhat more holistically than our sometimes more piecemeal efforts". The Hudson is an excellent candidate for such coordination. Present federal research on the system includes a Superfund project to reassess the Polychlorobiophenyl (PCB) problem (Environmental Protection Agency); the New York-New Jersey Harbor Estuary Program (Environmental Protection Agency); involvement in the National Oceanic and Atmospheric Agency (NOAA) National Status and Trends Program; studies of its tidal marshes as part of the National Estuarine Sanctuaries Program (NOAA); and U.S. Geological Survey (USGS) studies as part of their National Water Quality Assessment Program. Additional research funds, on the order of a million dollars a year, are supplied by a private foundation, the Hudson River Foundation. State and city efforts, policy and regulatory considerations, and active participation of environmental groups such as Natural Resources Defense Council and Environmental Defense Fund add to the challenge of arriving at a holistic approach. It seems that a logical initial step in this direction involves a consideration of research needs and directions. Past experience on the Hudson suggests the following:

BASIC INFORMATION

Under this heading, we include freshwater discharge, salinity distributions, and suspended-matter concentrations. These parameters provide the primary constraints on hydrodynamic models and form the basis for our understanding of contaminant transport (Figure 5.1) and nutrient dynamics. For the

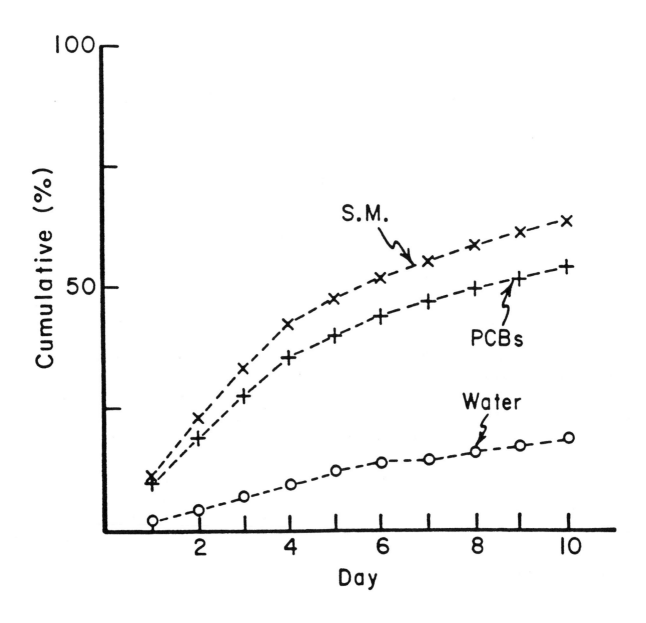

FIGURE 5.1 Cumulative percentage of PCB, suspended matter, and water transport from the upper Hudson for the ten days of highest suspended-matter concentration for the period from March 1, 1983 through September 30, 1983. Flow and suspended matter concentration data were from the USGS monitoring station at Stillwater, NY (USGS, 1984). PCB measurements and transport calculations were from Bopp et al. (1985). This plot emphasizes the importance of high-flow events in particle and associated contaminant transport.

Hudson, our concern focuses on the latter two parameters. To our knowledge, there is no continuous, long-term salinity-monitoring data for the Hudson estuary prior to 1987 when the USGS established two stations. The USGS has monitored suspended-matter concentrations daily since 1977 at the two stations, both located above the first dam on the river (Figure 5.2). Estimates of transport of suspended particles in the tidal Hudson basin generally refer to a data set collected at a single station during the 1960 water year Panuzio, 1965 and to limited data sets of Arnold (1982), Suszkowski (1978), the Institute for Ecosystem Studies (Findlay, 1991), and the Lamont-Doherty Geological Observatory (Olsen, 1979).

Continuous measurements of salinity (conductivity) and suspended particle concentrations (turbidity) are now fairly routine in oceanographic studies. Increased application of this type of monitoring in rivers and estuaries should be pursued.

TRENDS AND THE CURRENT SITUATION

Particle-Associated Contaminant Levels

A useful tool for studying trends in particle-associated contaminant levels is the analysis of dated sediment core sections (Alderton, 1985). Depositional timescales are generally determined from the depth distributions of natural (Pb-210, Be-7) and anthropogenic (Cs-137, Pu-239,240) radionuclides. Among the earliest applications of this technique to estuaries is the work of Goldberg on trace metal pollution (Goldberg, 1976; Goldberg et al., 1977, 1978, 1979). Continued and expanded use of this technique is a major emphasis of NOAA's National Status & Trends Program.

In the Hudson, analyses of dated sediment cores have helped to define the response time of the system to pollution events (Figure 5.3), to assess the contribution of contaminants from New York Metropolitan Area sewage treatment plants, and to quantify the improvements in levels of chlorinated hydrocarbons that were banned in the 1970s (Figure 5.4) (Bopp et al., 1982; Bopp and Simpson, 1989).

Recently, studies utilizing dated sediment core samples have expanded in scope, both in terms of the number of contaminants analyzed and the number of rivers and estuarine systems studied. We believe that this is a very positive development that should be encouraged. We strongly suggest that, in addition to time-trend analysis, other aspects of this technique should be pursued. For example:

The geographic distribution of current levels of particle-associated contaminants can be established through analysis of near-surface sediments in areas of rapid deposition. Such areas can be identified through measurements of the short-lived, particle-associated radionuclides such as Be-7 ($t_{1/2}$=53.4 days). Gram-sized samples of *current* particles suitable for contaminant analyses can also be collected by large-volume in situ filtration (Bishop et al., 1985), a technique developed for oceanographic studies that should be much more widely applied to river and estuarine systems.

Such studies can help to identify major point sources of contaminants (Figure 5.5) and provide an indication of the scale of particle and associated contaminant mixing and transport (Table 5.1).

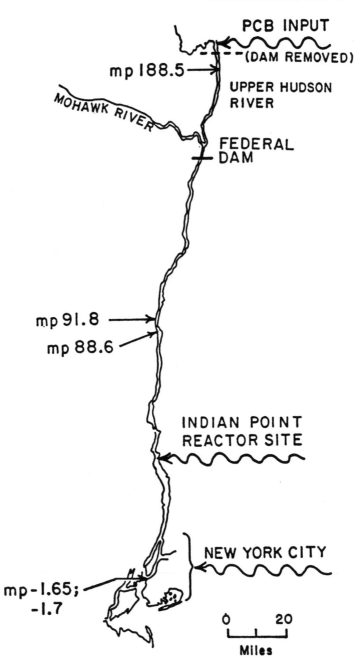

FIGURE 5.2 Map of the Hudson River. The two daily suspended-matter monitoring stations are located upstream of the Federal Dam. Locations of sediment core samples are given in mile points (mp), the number of statute miles upstream of the southern tip of Manhattan. (Source: Bopp and Simpson, 1989)

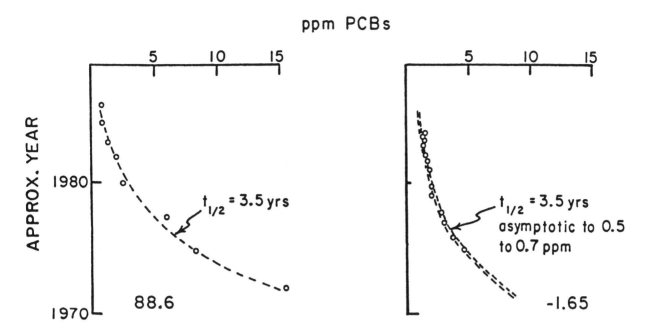

FIGURE 5.3 Data from dated cores from the Hudson River showing a dramatic decline in PCB levels in sediments deposited between the mid 1970's and the late 1980's. The data were discussed in greater detail. (Source: Bopp and Simpson, 1989)

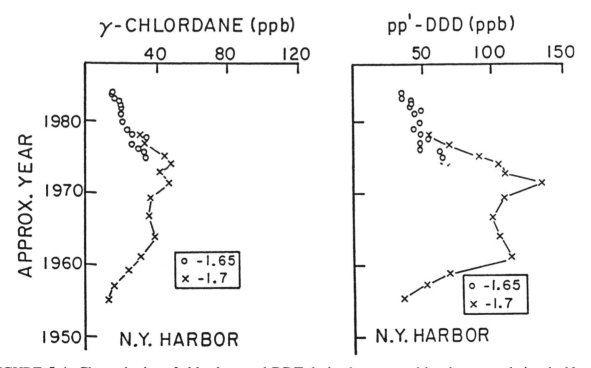

FIGURE 5.4 Chronologies of chlordane and DDT-derived compound levels accumulating in New York Harbor. (Source: Bopp and Simpson, 1989)

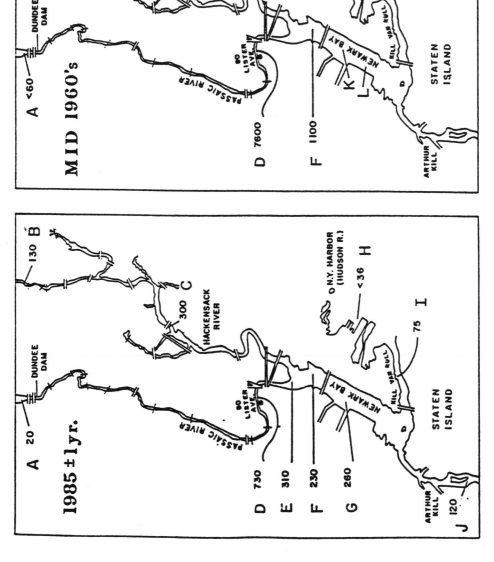

FIGURE 5.5 Levels of 2,3,7,8-TCDD (parts per trillion) on particulars from two horizons from Newark Bay and contiguous water. The 1985 ±1 year samples were suspended matter (sites G & J) and core top samples from areas of rapid deposition characterized by significant activity of the short-lived radionuclide, Be-7. The mid-1960s time horizon samples were chosen based on the depth profile of the fallout radionuclide, Cs-137. The data are consistent with a major source of 2,3,7,8-TCDD from a chemical manufacturing facility at 80 Lister Avenue on the lower Passaic River. (Source: Bopp et al., 1991a) Reprinted with permission from Americal Society, 1991.

TABLE 5.1 Levels of chlorinated hydrocarbons on particles from two time horizons from New York Harbor. The two sets of samples were core tops collected during 1982-84 and 1988-89, characterized by significant activity of the short-lived radionuclide, Be-7. The data indicates that levels of these contaminants are similar throughout the Harbor and that PCB and pp'-DDD levels continue to decline.

	PCBs	ppb pp'-DDD	$\propto$-chl
1982-84			
-1.65	1470	35	15
-0.8	1930	47	19
2.3	1930	36	16
2.35	1410	36	12
2.75	1220	32	11
1988-89			
-1.68	830	24	11
2.34	770	21	12
2.73	850	33	14
6.3	740	21	21
9.7	720	33	14

(Source: Bopp et al., 1991b)

Levels of particle-associated contaminants, combined with a knowledge of the time of deposition or transport, can be a powerful tracer of net particle transport. This is most important in estuaries where direct determination of net particle transport is extremely difficult, since it generally is a relatively small difference between two large numbers, downstream transport with ebbing currents and upstream with the flood. Some examples from Newark Bay, New Jersey, are shown in Figure 5.6 and Table 5.2.

Pairs of dated sediment cores, collected several years apart from the same site, can be used as in situ incubation experiments to study the degradation rates and pathways of organic contaminants. The discovery of anaerobic dechlorination of PCBs in the upper Hudson River (Brown et al., 1984) has focused significant scientific attention on in situ degradation. Figure 5.7 shows an example of a pair of sediment cores that could be used to study this process.

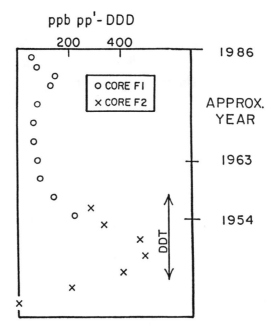

FIGURE 5.6 A chronology of pp'-DDD levels in Newark Bay sediments. Maximum levels in the late 1940s to early 1950s have been associated with the operation of a DDT manufacturing facility. This input could provide an excellent tracer of particle and associated contaminant movement in this estuary. (Source: Bopp et al., 1991)

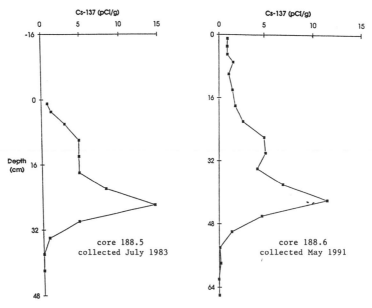

FIGURE 5.7 Depth profiles of Cs-137 activities in sediment cores collected at the same site in the upper Hudson River. The profiles indicate that samples from the two cores can be paired off. The same samples from the more recent core will represent an eight-year in situ incubation. Such samples could be quite useful in studies of in situ degradation of organic contaminants. Reprinted with permission from American Chemical Society, 1991.

TABLE 5.2 Levels of pp'-DDD and -chlordane on suspended particle samples from Newark Bay near the downstream end of the Hackensack River and the upper Hackensack River. The distinctive signatures suggest significant potential for tracing particles and associated contaminant mixing in the tidal reach of the Hackensack.

	pp'-DDD (ppb)	∝-chlordane (ppb)	pp'-DDD / ∝-chl
NEWARK BAY			
F1049 4/26/85	112	35	3.2
F1054 8/20/85	110	29	3.8
F1077	103	37	2.8
Upper Hackensack River			
F1051 4/26/85	61	116	0.53
F1052 7/30/85	48	111	0.43
F1075 3/11/86	55	84	0.65

We strongly suggest the archiving of sediment core samples for application in this type of study and for future use in other investigations.

Nutrients

There is a need for greater commitment to long-term, research-based monitoring and coordination of efforts. For the Hudson, agencies including the Interstate Sanitation Commission and the New York City Department of Environmental Conservation have collected historical monitoring data. Researchers at the Lamont-Doherty Geological Observatory and the Institute for Ecosystems Studies have worked on nutrient dynamics. The USGS National Water Quality Assessment program on the Hudson will include nutrient monitoring, and the Harbor Estuary Program is considering historical nutrient data and future research directions.

Wastewater discharge plays a major role in determining the levels of nutrients in Hudson estuary waters. Major impacts on the nutrient dynamics of the estuary would be expected as the result of New York's ban on phosphates in detergents and improvements in sewage treatment over the past two decades. Figure 5.8 summarizes what we consider to be the best data available for quantifying the

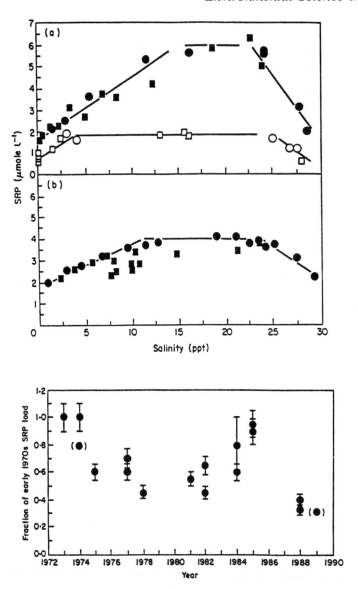

FIGURE 5.8 Plots of soluble reactive phosphate (SRP) versus salinity for the lower Hudson River. Many transects show the major input at mid salinities associated with New York Metropolitan Area wastewater discharge and quasi-conservative mixing with coastal ocean and fresh water endmembers. Panel "a" is from the mid-1970s. Solid symbols represent a period of low fresh water discharge (summer). Open symbols show the effects of high spring discharge flushing the estuary. Panel "b" represents a low-flow transect from 1988. The lower peak levels of SRP compared with those in the late 1970s reflect the effect of New York's ban on phosphate-based detergents and upgrades in Metropolitan Area sewage treatment. The lower panel shows the history of SRP loading to New York Harbor, derived from modeling similar transects. The general decrease in loading with time is interrupted by higher loadings in the mid-1980s that could be associated with reconstruction of primary clarifiers at a major sewage treatment plant. (Source: Clark et al., 1992a) Reprinted with permission from Academic Press Ltd, London, England, 1992.

changes in phosphate levels. The 15 years of collecting data were accomplished not as a result of any long-term funding initiative but rather through the creative use of piecemeal funding. This situation underscores the need for long-term funding commitments and coordination of efforts.

RESEARCH AND REGULATION-DRIVEN MONITORING

There seems to be significant and largely untapped potential for collaboration between research scientists and agencies involved in monitoring programs. For example, the New York City Department of Environmental Protection has regular cruises on the lower Hudson for the purpose of collecting water column and sediment-monitoring data. Participation of research scientists would make more efficient use of the ship time, broaden the data base, and help with data interpretation. For New York Harbor, sewage-treatment-plant effluent is an important source of contaminants, including trace metals and PCBs. Despite national focus on the Hudson River PCB problem, monitoring efforts have produced little useful quantitative data on the magnitude of this PCB source. A research-based approach, with modifications for the sampling, cleanup, and analytical procedures, could supply much-needed information on a limited set of samples. By necessity, this would be a joint effort involving cooperation between research scientists and the regulatory agencies.

Future use of filled wetlands bordering rivers and estuaries is an area with great potential for combining research and regulation-driven monitoring. The metropolitan region at the terminus of the Hudson is situated above groundwater aquifers of the Atlantic Coastal Plain and of more recent glacial origin. The character of the area is strongly urban and land usage has led to degradation of groundwater resources. An important source of pollutants in these aquifers is landfills. Formal strategies of wetland filling with solid waste have resulted in the eradication of most of the wetlands in the lower Hudson region, including over 45,000 acres in New York City (Walsh, 1991a). These landfills cover over 20 percent of the city's land area (Figure 5.9, Walsh, 1991b). Competition for real estate is great in this area and old landfills are in demand because they are often the only remaining large tracts of undeveloped land. Government regulations that require intensive environmental testing of such sites prior to development have led to numerous studies that yield information on aquifer geology and the quality of water, soil, and sediment. For example, environmental studies have recently been performed on eight lots on the north shore of Jamaica Bay ranging in size from 4 to 297 acres. The lots were part of a 6000-acre area of former tidal wetlands that were filled with solid wastes between 1910 and 1980 (Walsh, 1991a). The studies were all performed independently and, because of the narrow focus of each, a valuable opportunity to gain insight into the hydrologic and geochemical dynamics of this expansive landfill was lost. The importance of groundwater resources in densely populated coastal environments necessitates more effective coordination of future site investigations. If sufficient flexibility can be accommodated within the regulatory requirements, such investigations could form the basis of an important research effort on landfills and coastal aquifers.

FIGURE 5.9 A map showing landfilled areas in New York City.

OTHER SPECIFIC RESEARCH AREAS

Gas Exchange

Models of atmospheric oxygen inputs and nitrous oxide and organic contaminant export from estuaries generally relate turbulence to wind speed (Emerson, 1975), tidal flow (O'Connor and Dobbins, 1958), or the relatively few direct measurements that have been made. Recent work in lakes using injected tracers (Wanninkhof et al., 1985) shows great promise for direct determination of gas exchange. In the Hudson, injection of freons with sewage-treatment-plant effluent may provide a very useful gas-exchange tracer (Clark et al., 1992b; Figure 5.10).

Fine-Particle Behavior in the Coastal Environment

What is the fate of fine particles and associated contaminants flushed from estuaries during high-discharge events? What fraction is deposited on the shelf? Is there significant transport to the deep ocean or onshore to estuaries and coastal bays? To what extent do estuaries act as particle traps? These basic and very significant questions are amenable to study with geochemical tracers. In the Savannah River estuary, (Olsen et al., 1989) utilized plutonium isotopic ratios and C-13 abundance in organic matter to study the importance of onshore particle transport, elaborating on an earlier study of Goldberg et al. (1979). In Baltimore Harbor, zinc and other trace elements were found to be effectively trapped by sedimentation, despite rapid flushing of the waters (Sinex and Helz, 1982). Some recent data on DDT-derived compounds in the coastal environment suggest that atmospheric inputs of unaltered DDT may provide a most useful tracer (Bopp et al., 1993; Table 5.3 and Figure 5.11).

CONCLUSIONS

Our observations on the Hudson indicate the need for attention to the basic monitoring data base, demonstrate that creative use of geochemical tracers should play an important role in future investigations, and suggest that coordination of research and monitoring efforts and continuity of research funding should be a high priority.

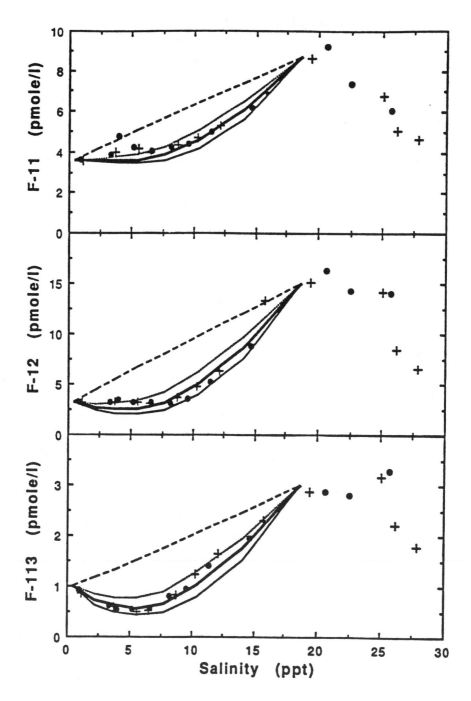

FIGURE 5.10 Levels of freons dissolved in Hudson River water plotted versus salinity. The mid-salinity peak has been associated with inputs from New York Metropolitan Area wastewater discharges into New York Harbor. The dashed line shows concentrations expected for conservative mixing between the harbor and the fresh water endmember. The data falls between model generated lines that allow for gas exchange at piston velocities of 1 and 4 cm/sec. (Source: Modified from Clark et al., 1992b)

TABLE 5.3 pp'-DDT to pp'-DDD ratios in samples from the Hudson estuary and samples from the continental slope, about 100 miles offshore. Lower rates in the estuary samples reflect anaerobic conversion of DDT to DDD. Much higher ratios in the offshore samples have been interpreted as the result of atmospheric inputs of fresh DDT.

	$\dfrac{\text{pp'-DDT}}{\text{pp'DDD}}$
ESTUARY SAMPLES	
Suspended Particles	0.11-0.63
Sediment Cores	0.05-0.87
Sewage Samples	0.2-0.3
SLOPE SAMPLES	
Aerobic Sediments	13-24
Trap Samples	1.6-5.4

REFERENCES

Alderton, D. H. M. 1985. Sediments, p. 1-95 in Historical Monitoring, MARC Report Number 31, Monitoring and Assessment Research Centre, London, U.K., 320 pp.

Arnold, C. L. 1982. Modes of fine-grained suspended sediment occurrence in the Hudson River Estuary. M.S. Thesis, State University of New York, Stony Brook, New York.

Bishop, J. K. B., D. Schupack, R. M. Sherrell, and M. Conte. 1985. A multiple-unit large-volume in situ filtration system for sampling oceanic particulate matter in mesoscale environments, p. 155-175 in A. Zirino (ed), Mapping Strategies in Chemical Oceanography, American Chemical Society, Advances in Chemistry 209.

Bopp, R. F., H. J.Simpson, C. R. Olsen, R. M. Trier, and N. Kostyk. 1982. Chlorinated hydro-carbons and radionuclide chronologies in sediments of the Hudson River and estuary, New York, Environ. Sci. Technol., v. 16, p. 666-676

Bopp, R. F., H. J. Simpson, and B. L. Deck. 1985. Release of Polychlorinated Biphenyls from Contaminated Hudson River Sediments, Final Report NYS C00708, New York State Department of Environmental Conservation, Albany, NY, 89 pp.

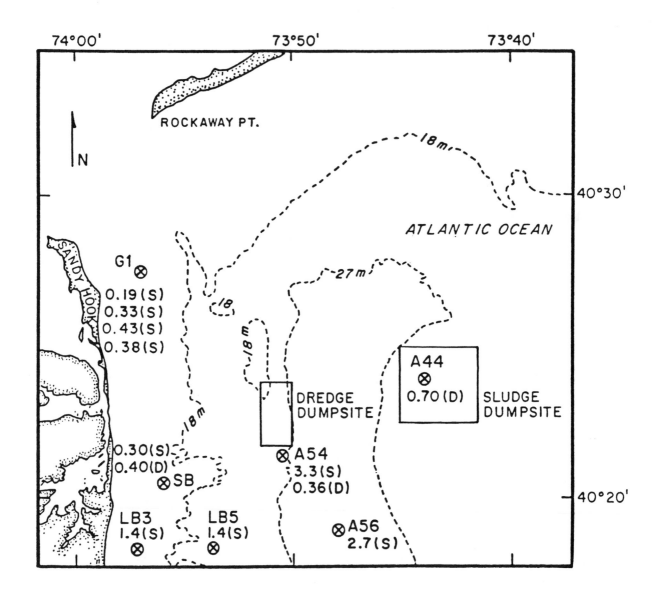

FIGURE 5.11 pp'-DDD to pp'-DDD ratios in surface (S) and deep (D) suspended particles from the coastal ocean near the mouth of the Hudson. Samples with ratios less than unity primarily reflect the influence of Hudson discharge (sites G1 and SB) or resuspension of bottom sediments (A54(D)). Samples with ratios greater than one suggest significant influence of atmospheric inputs of unaltered DDT. (Source: Bopp et al., 1993)

Bopp, R. F., and H. J. Simpson. 1989. Contamination of the Hudson River: the sediment record, p. 401-416 in Contaminated Marine Sediments: Assessment and Remediation, Washington, D.C.: National Academy Press.

Bopp, R. F., M. L. Gross, H. Tong, H. J. Simpson, B. L. Deck, and F. C. Moser. 1991a. A major incident of dioxin contamination: sediments of New Jersey estuaries, Environ. Sci. Technol., v. 25, p. 951-956.

Bopp, R. F., H. J. Simpson, D. W. Robinson, S. N. Chillrud, and A. Virgilio. 1991b. Chronologies of persistent contaminants in New York Harbor, Raritan Bay and Jamaica Bay, Abstract, Estuarine Research Federation Conference, San Francisco, CA, Nov. 10-14, 1991.

Bopp, R. F., D. W. Robinson, H. J. Simpson, P. E. Biscaye, R. F. Anderson, H. Tong, S. J. Monson, and M. L. Gross. 1993. Recent sediment and contaminant distributions in the Hudson Shelf Valley, 12-MDS Symposium Volume, NOAA Tech. Mem., in press. Ann Studholme, Editor. NOAA National Marine Fisheries in Sandy Hook, NJ.

Brown, J. F. Jr., R. E. Wagner, D. L. Beddard, M. J. Brennan, J. C. Carnahan, and R. J. May. 1984. PCB transformations in upper Hudson sediments, Northeastern Env. Sci. 3:167-179.

Clark, J. F., H. J. Simpson, R. F. Bopp, and B. L. Deck. 1992a. Geochemistry and loading history of phosphate and silicate in the Hudson estuary, Estuarine, Coastal and Shelf Science 34:213-233.

Clark, J. F., H. J. Simpson, W. M. Smethie, Jr., and C. Toles. 1992b. Gas exchange in a contaminated estuary inferred from chlorofluorocarbons, Geophys. Res. Lett. 19:1133-1136.

Emerson, S. 1975. Gas exchange rates in small Canadian sheild lakes, Limnol. Oceanogr. 20:754-761.

Findlay, S., M. Pace, and D. Lints. 1991. Variability and transport of suspended sediment, particulate and dissolved organic carbon in the tidal and freshwater Hudson River, Biogeochemistry 12:149-169.

Goldberg, E. D. 1976. Pollution history of estuarine sediments, Oceanus 19:18-26.

Goldberg, E. D., E. Gamble, J. J. Griffin, and M. Koide. 1977. Pollution history of Narragansett Bay as recorded in its sediments, Estuarine Coastal Mar. Sci. 5:549-561.

Goldberg, E. D., V. Hodge, M. Koide, J. Griffin, E. Gamble, O. P. Bricker, G. Matisoff, G. R. Holdren, Jr., and R. Braun. 1978. A Pollution History of Chesapeake Bay, Geochim. Cosmochim. Acta, 42:1413-1425.

Goldberg, E. D., V. F. Hodge, J. J. Griffin, M. Koide, and H. Windom. 1979. Pollution history of the Savannah River estuary, Environ. Sci. Technol. 13:588-594.

O'Connor, D. J., and W. E. Dobbins. 1958. Mechanism of reaeration in natural steams, Am. Soc. Civ. Eng., 123:641-694.

Olsen, C. R. 1979. Radionuclides, sedimentation and the accumulation of pollutants in the Hudson estuary. Ph.D. thesis, Columbia University, New York, NY.

Olsen, C. R., M. Thein, I. L. Larsen, P. D. Lowry, P. J. Mulholland, N. H. Cutshall, J. T. Bird and H. L. Windom. 1989. Plutonium, lead-210, and carbon isotopes in the Savannah estuary: riverborne versus marine sources, Environ. Sci. Technol. 23:1475-1481.

Panuzio, F. L. 1965. Lower Hudson River siltation, p. 512-560 in Proceedings of the Federal InterAgency Sedimentation Conference, 1963; Agricultural Research Service Misc. Publ. No. 970.

Sinex, S. A., and G. R. Helz. 1982. Entrapment of zinc and other trace elements in a rapidly flushed industrialized harbor, Environ. Sci. Technol., 16:820-825.

Suszkowski, D. J. 1978. Sedimentology of Newark Bay, New Jersey: an urban estuary. Ph.D. Thesis, University of Delaware, Newark, Delaware.

U.S.Geological Survey. 1984. Water Resources Data for New York, Water Year 1983, USGS Water Data Report NY-83-1. Washington, D.C: U.S. Government Printing Office.

Walsh, D. C. 1991a. The history of waste landfilling in New York City, Ground Water 29:591-3.

Walsh, D. C. 1991b. Reconnaissance mapping of land fills in New York City, in Proceedings of FOCUS Conference on Eastern Regional Ground Water Issues, National Ground Water Association, Portland, ME, October 29-31, 1991.

Wanninkhof, R., J. R. Ledwell, and W. S. Broecker. 1985. Gas exchange-wind speed relation measured with sulfur hexafluoride on a lake, Science 27:1224-1226.

6

Types of Coastal Zones:
Similarities and Differences

Douglas L. Inman
Scripps Institution of Oceanography
La Jolla, California

INTRODUCTION

Coastal and estuarine waters are the parts of the sea that overwhelmingly dominate our everyday affairs. Our rapidly expanding use of the ocean, increasing excursion upon it, and entry into it are mostly concerned with processes that take place in shallow water. As well, it is mostly within coastal waters that human acts, such as waste discharge, fishing, dredging, mining, drilling, and coastal structures, have their greatest impact on the ocean. Accordingly, coastal waters and the underlying submerged lands are the areas of highest scientific interest and jurisdictional controversy.

This paper provides an overview of the world's five types of coastal zones. Their tectonic origins and shaping processes are compared and contrasted. An understanding of these different types of coasts and their nearshore processes is essential to policy-making efforts.

FACTORS DETERMINING COASTAL ZONE TYPES

The common types of coastal zone are well represented along the shores of the United States. These types range from the ice-push coasts of Alaska to the coral reef coasts of Hawaii and southern Florida. They include, as well, the far more common types, such as the barrier beach coasts of the Atlantic; the steep, cliff-backed coasts of the Pacific; and the marginal-seas-type coast of the Gulf of Mexico. Although there are general processes that apply to all coasts, there are also significant differences among coastal types.

These similarities and differences stem from the influence of various processes. The most important of these processes are

- tectonics;
- exposure to waves, winds, and ocean currents;
- tidal range and intensity of current;

- supply of sediment and its transport along the coast; and
- coastal climate.

The position and configuration of the continental shelf and adjacent coast are related to the moving tectonic plates. This geologic setting and the exposure to waves are the two most significant factors in determining nearshore processes. Waves, winds, and currents are the principal driving forces for coastal processes, and have extensively modified the coast by the erosion and deposition of sediment. Coastal climate is mainly dependent upon latitude and the location of the major ocean and atmospheric current systems. Extremes in coastal climate associated with latitude result in the unique aspects characteristic of arctic coasts in the north and coral reef coasts near the equator.

The tectonic and paleoclimatic processes important to the geologic setting of coasts operate over the largest areas and have the longest time scales. Since the large-scale features of a coast are associated with its position relative to plate margins, plate tectonics provides a convenient basis for the first order classification of coasts, that is, longshore dimensions of about 1000 km (Inman and Nordstrom, 1971). Such a classification leads to the definition of three general tectonic types of coasts: (1) collision coasts, (2) trailing-edge coasts, and (3) marginal sea coasts.

Collision coasts are those that occur along active plate margins, where the two plates are in collision or impinging upon each other (Figure 6.1). Tectonically, this is an area of crustal compression and consumption. These coasts are characterized by narrow continental shelves bordered by deep basins and ocean trenches. Submarine canyons cut across the narrow shelves and enter deep water. The shore is often rugged and backed by sea cliffs and coastal mountain ranges; earthquakes and volcanism are common. The sea cliffs and mountains often contain elevated sea terraces representing former relations between the level of the sea and the land (Figure 6.2). The west coasts of South and Central America are typical examples of collision coasts. Although much of the California coast is now a northward-moving terrain associated with the San Andreas fault, this coast retains most of the characteristics of its collision history.

Trailing-edge coasts occur on the *trailing-edge* of a land mass that moves with the plate. They are thus situated upon passive continental margins that form the stable portion of the plate, well away from the plate margins. The east coasts of North and South America are examples of mature, trailing-edge coasts. These coasts typically have broad continental shelves that slope into deeper water without a bordering trench. The coastal plain is also typically wide and low-lying and usually contains lagoons and barrier islands, as on the east coasts of the Americas (Figure 6.2).

Marginal sea coasts are those that develop along the shores of seas enclosed by continents and island arcs. Except for the Mediterranean Sea, these coasts do not usually occur along plate margins since the spreading center margins are commonly in ocean basins, while the collision edges of plates face oceans. These coasts are typically bordered by wide shelves and shallow seas with irregular shorelines. The coastal plains of marginal sea coasts vary in width and may be bordered by hills and low mountains. Rivers entering the sea along marginal sea coasts often develop extensive deltas because of the reduced intensity of wave action associated with small bodies of water. Typical marginal sea coasts border the South and East China Seas, the Sea of Okhotsk, and the Gulf of Mexico.

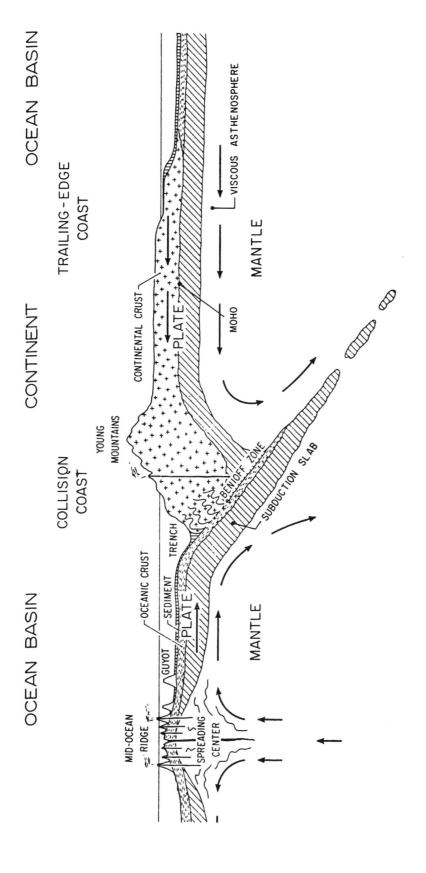

FIGURE 6.1 Schematic illustration of the formation of a collision coast and a trailing-edge coast. Representative of a section from the East Pacific Rise (spreading center) through the Peru-Chile trench off South America at 34 degrees south latitude. (Source: Inman and Nordstrom, 1971). Reprinted with permission from The University of Chicago Press, 1991.

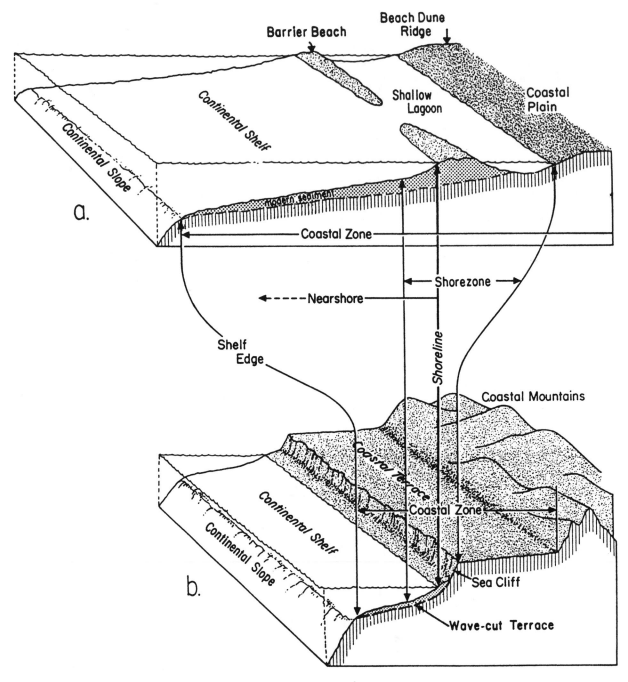

FIGURE 6.2 Definition sketch for coastal zone nomenclature. The type of coast is related to its relative position on the moving plates of the tectosphere; wide-shelf plains coasts (a) and narrow-shelf mountainous coasts (b) are characteristic of the east coast (trailing edge) and west coast (collision edge) of the Americas, respectively. (Source: Inman and Brush, 1973) Reprinted with permission from Science, 1973.

It is apparent that the morphologic counterparts of collision coasts, trailing- edge coasts, and marginal sea coasts become respectively: narrow-shelf hilly and mountainous coasts, wide-shelf plains coasts, and wide-shelf hilly coasts. A complete classification would also include coasts formed by other agents such as glacial scour, ice-push, and reef-building organisms, adding two other types of coast: cryogenic coasts and biogenic coasts. Common examples of the latter two coastal types described here are arctic coasts and coral reef coasts.

Paleoclimate and Sea-level Change

Climate, through its control of glaciation, is the principal factor leading to changes in sea level. The Pleistocene Epoch is characterized by cycles of alternate cold and warm periods producing glacial and interglacial stages.

The last glacial stage, known as the Wisconsinan, had a maximum about 18,000 years BP. Since that time, the climate has warmed causing glaciers to melt and sea level to rise in what is generally known as the Flandrian transgression (Figure 6.3). Tide gauge records indicate that sea level is still rising on a worldwide (eustatic) basis at a rate of about 15 cm per century (Barnett, 1984; National Research Council, 1987), and there is the distinct possibility of an increased rate of rise due to the greenhouse effect of carbon dioxide released by man in coming years (e.g., Emery, 1980). This continuing rise in sea level increases sea cliff erosion and produces a gradual retreat of beaches and barrier islands on a worldwide scale. If all of the ice on earth were to melt, it would raise sea level about 78 meters above present level (Barry, 1981).

Sea-level curves for deglaciated areas show a net emergence due to the *glacial rebound* associated with the removal of the ice load (Figure 6.4). Viscoelastic models (e.g., Peltier, 1986) show that uplift occurs in the areas of greatest ice loading and that a drawdown (subsidence) can occur in areas marginal to the area of loading. This may explain why portions of the mid-Atlantic coast of the United States show relative sea-level rise of about 30 cm/century; one-half of this may be due to eustatic sea-level rise while the remainder is viscoelastic draw-down (Figure 6.4).

The present, relatively long, near stasis in sea level has produced coastlines that are unique for the Holocene and probably for the entire Pleistocene Epoch. The sea level has been relatively high during the past three thousand to six thousand years, accentuating the broad shelves carved into the continental platform during this and previous high stands. As a consequence, stream valleys cut at lower sea level are filling; streams near the coast are *at grade;* and coastlines typically have long, continuous beaches of sand.

COASTAL PROCESSES

Similarities and differences in coastal types are most easily understood in terms of *nearshore circulation cells* and the budget of sediment in *littoral cells.* Nearshore circulation cells determine the path of wave-driven water circulation on a local scale of about 1 km on ocean beaches, while the

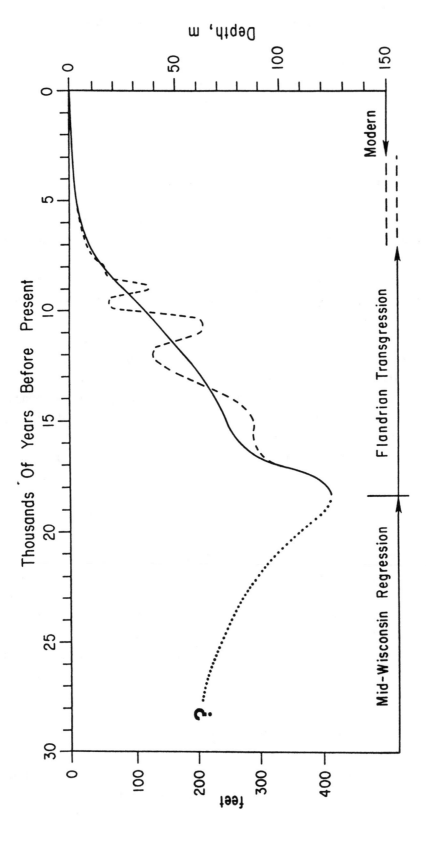

FIGURE 6.3 Late Quaternary fluctuations in sea level. Solid line is the *generalized* sea level curve (from Curray, 1965); dashed line is detailed curve (from Curray, 1960, 1961). Tree ring and Uranium/Thorium dates give greater age than the radiocarbon ages for these curves. Recent studies (Fairbanks, 1989; Bard, 1990) indicate the glacial maximum was 21,000 (^{230}TH/^{234}U) years BP with a sea level lowering of 121 ± 5 m.

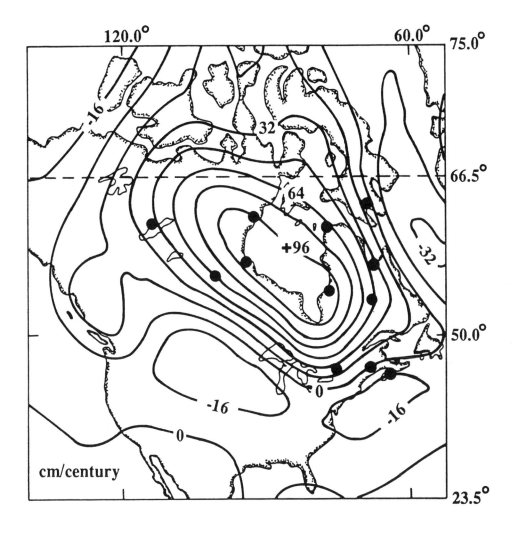

FIGURE 6.4 The predicated rate of uplift (+) and subsidence (-) in cm/century resulting from Laurentide deglaciation according to the viscoelastic model of Peltier (1986).

budget of sediment concerns the sources, transport paths, and sinks of sediment in a littoral cell of coastal length 10 km to 100 km.

Nearshore Circulation

The interaction of surface waves moving toward the beach with other, trapped waves traveling along the shore produces alternate zones of high and low waves that determine the position of seaward-flowing rip currents. The rip currents are the seaward return flow for the longshore currents that flow parallel to the shore inside of the surf zone. The pattern that results from this flow takes the form of a horizontal eddy or cell, called the nearshore circulation cell (Inman et al., 1971, Figure 6.5). Nearshore circulation cells are ubiquitous wherever waves break along sandy beaches, and the intense, concentrated, seaward flow of their rip currents is the principal cause of drowning for inexperienced swimmers.

The nearshore circulation system produces a continuous interchange between the waters of the surf zone and the shelf, acting as a distributing mechanism for nutrients and as a dispersing mechanism for land runoff. Offshore water is transported into the surf zone by breaking waves, and particulate matter is filtered out on the sands of the beach face. Runoff from land and pollutants introduced into the surf zone are carried along the shore and mixed with the offshore waters by the seaward-flowing rip currents.

Two important mixing mechanisms are operative within the surf zone, each having distinctive length and time, scales determined by the intensity of the waves and the dimensions of the surf zone. The first is associated with the breaking wave and its bore, which produce rapid mixing in an onshore--offshore direction. This mixing gives coefficients of eddy diffusivity of the order of $H_b X_b/T$ where H_b, and T are the breaker height and period of the waves and X_b is the width of the surf zone. The second process is advective and is associated with the longshore and rip current systems in the nearshore circulation cell. This longshore mixing mechanism gives an apparent eddy-mixing coefficient of the order of Yv_ℓ, where v_ℓ is the longshore current velocity and Y is the longshore spacing between rip currents. Along ocean beaches $H_b X_b/T$ and Yv_ℓ are about 10 m²/s and 100 m²/s respectively (Inman et al., 1971).

In addition, coastal circulation cells of large dimension are associated with the submarine canyons that cut across the shallow shelves of the world (Inman et al., 1976). The submarine canyons act as deep, narrow conduits connecting the shallow waters of the shelf with deeper water offshore. At times, strong seaward flows of water occur in the canyons, resembling large-scale rip currents. The canyon currents produce circulation cells having the dimensions of the shelf width and the spacing between the submarine canyons. These strong currents in submarine canyons seem to be caused by a unique combination of air-sea-land interactions consisting of: (1) a *pile-up* of water along the shoreline caused by strong onshore winds, (2) down-canyon pulses of water caused by the alternate high and low grouping (surf beat) of the incident waves, (3) a shelf seiche excited by the waves and by the pressure fluctuations in the wind field, and (4) the formation of continuous down-canyon

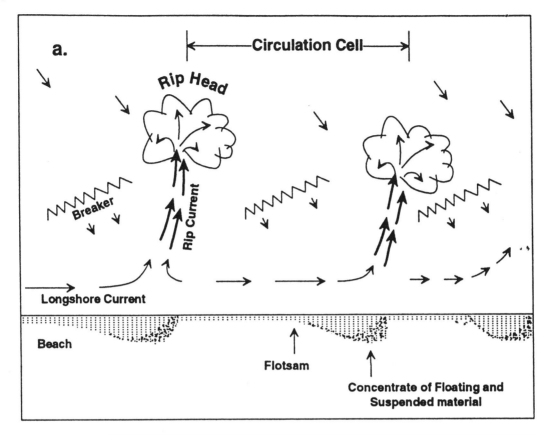

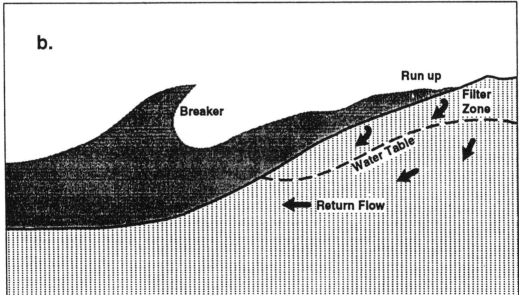

FIGURE 6.5 Schematic diagram of nearshore circulation cell consisting of onshore transport by the breaking wave, longshore transport in the surf zone and offshore transport by seaward flowing rip currents. Floating and suspended material is deposited on the beach face by (a) wave runup and by (b) water percolating through the beach sand.

currents as the accumulated weight of the sediment dislodged by the currents overcomes the density stratification of the deeper water.

Littoral Cells and the Budget of Sediment

A basic approach to understanding the relative importance of nearshore processes is to compare the sea's potential to erode the land with the land's potential to supply terrestrial erosion products. Such a comparison ultimately resolves itself in the balance between the budget of power in waves and currents and the budget of sediments available for transport. Of course, this balance varies widely from place to place and, even in the best studied areas, is but poorly understood. However, order of magnitude estimates can be attempted by considering the types of driving forces and the resulting sediment response in terms of the budget of sediment.

Waves move sand on, off, and along the shore. Once an equilibrium beach profile is established, the principal transport is along the coast. Theory and measurements show that the longshore transport rate of sand is proportional to the longshore stress-flux of the waves.[1]

The budget of sediment for a region is obtained by assessing the sedimentary contributions and losses to the region and their relation to the various sediment sources and transport mechanisms. Determination of the budget of sediment is not a simple matter, since it requires knowledge of the rates of erosion and deposition as well as understanding of the capacity of various transport agents. Studies of the budget of sediment show that coastal areas can be divided into a series of discrete sedimentation compartments called *littoral* cells. Each cell contains a complete cycle of littoral transportation and sedimentation including transport paths and sources and sinks of sediment. Littoral cells take a variety of forms, but there are two basic types. One is characteristic of collision coasts with submarine canyons, while the other is more typical of trailing-edge coasts where rivers empty into large estuaries as shown in Figure 6.6 (e.g., Inman and Brush, 1973; Inman and Dolan, 1989).

COASTAL ZONE SIMILARITIES AND DIFFERENCES

Mixing and longshore transport of nutrients, pollutants, and sediment occur in the nearshore circulation cells that are ubiquitous to all coastal zones. However, the dimensions and intensities of mixing and sediment transport are determined principally by wave climate. Higher waves produce wider surf zones and more intense mixing and transport. In general, windward coasts like the Pacific coast of the United States are subject to more consistent wave action with seasonal variations in intensity between summer and winter. In contrast, leeward coasts like the mid-Atlantic coast of the United States tend to have lower levels of average wave intensity but episodic interruptions by occasional severe tropical storms in summer and extratropical *northeasters* in winter (e.g., Inman and

---

[1]The longshore stress-flux is CS_{YX} where C is the phase velocity of the nearbreaking waves and S_{YX} is the longshore radiation stress (e.g., Inman and Brush, 1973; Inman and Dolan, 1989; equations 6.2, 6.3).

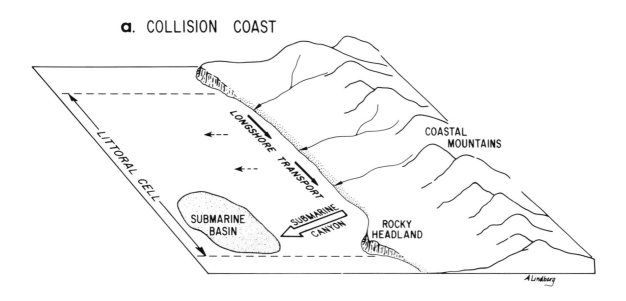

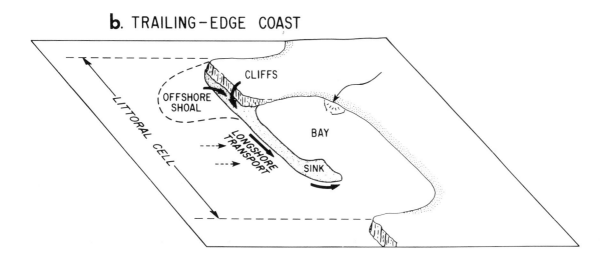

FIGURE 6.6 Sediment source, transport paths, and sinks for typical littoral cells along (a) collison and (b) trailing-edge coasts. Arrows show sediment transport paths; dotted arrows indicate occasional onshore and offshore transport modes.

Dolan, 1989). This results in more consistent mixing and transport processes on windward coasts and more episodic processes on leeward coasts.

The elements in the budget of sediment may be significantly different among the coastal types. These differences are associated with the steepness of the continental shelf and with the proximity of coastal mountains and streams that debouch directly into the sea. For example, rivers and streams are generally important sediment sources for collision coasts, whereas shelf and barrier roll-over are generally more important sources along trailing-edge coasts.

Collision Coasts

Collision coasts are erosional features characterized by narrow shelves and beaches backed by wave-cut seacliffs. Along these coasts with their precipitous shelves and submarine canyons, as in California, the principal sources of sediment for each littoral cell were the rivers, which periodically supplied large quantities of sandy material to the coast. The sand is transported along the coast by waves and currents until the *river of sand* is intercepted by a submarine canyon, which diverts and channels the flow of sand into the adjacent submarine basins and depressions (Figure 6.6a).

In the San Diego region of California, most coastal rivers have dams that trap and retain their sand supply. Studies show that in this area the yield of sediment from small streams and coastal blufflands has become a significant new source of sediment. It was also found that the cluster storms associated with the 1982/83 El Niño-Southern Oscillation phenomena produced beach disequilibrium that resulted in downwelling currents that carried sand onto the shelf (Inman and Masters, 1991). Normal wave action contains sand against the coast and, when sediment sources are available, results in accretion of the shorezone. High, total-energy wave events cause a loss of sand from the shorezone via downwelling currents that deposit sand on the shelf. The downwelled sediment is lost to the shorezone when deposited on a steep shelf such as that off Oceanside, California (Figure 6.6a), or it may be returned gradually from a more gently sloping shelf to the shorezone by wave action. The critical value of slope for onshore transport of sand varies with depth and wave climate but for depths of about 15 to 20 m is about 1.5 percent (1.0 degree).

In all cases where measurements were made just before and after the 1982/83 cluster storm events and the profiles were distant from structures, it was found that these storm events resulted in the lowest level of beach sand in the history of the observations. Using the profiles north of Oceanside Harbor, where conditions are closest to natural and unaffected by harbor effects, it was found that the 38 km of the central Oceanside subcell during 1982/1983 lost an unprecedented 33 million cubic meters of sand from the shorezone! Such a volume represents perhaps a 50-year supply of sediment to the shorezone under normal conditions (Inman and Masters, 1991).

California beaches are narrow and backed by eroding seacliffs that in many places have buildings on their brink. Since a wide beach is the best protection for eroding seacliffs, a major problem for these coasts is finding adequate sources of sand for beach nourishment.

Trailing-Edge and Marginal Sea Coasts

The mid-Atlantic coast of the United States, with its characteristic wide shelf bordered by coastal plains, is a typical trailing-edge coast. This low-lying barrier island coast has large estuaries occupying drowned river valleys. River sand is trapped in the estuaries and cannot reach the open coast. For these coasts, the sediment source is from beach erosion and shelf sediments deposited at a lower stand of the sea, whereas the sinks are sand deposits that tend to close and fill the estuaries (Figure 6.6b). Under the influence of a rise in relative sea level, the barriers are actively migrating landward in a rollover process in which the volume of beach face erosion is balanced by rates of overwash and fill from migrating inlets (e.g., Leatherman, 1979, 1981; Inman and Dolan, 1989). For these coasts, the combination of longshore transport and rollover processes leads to a distinctively *braided* form for the *river of sand*.

The Outer Banks of North Carolina, which include the Hatteras and Ocracoke Littoral Cells, extend for 320 kilometers and are the largest barrier island chain in the world. The Outer Banks are barrier islands separating Pamlico, Albemarle, and Currituck Sounds from the Atlantic Ocean. These barriers are transgressing landward, with average rates of shoreline recession of 1.4 m/yr. between False Cape and Cape Hatteras. Oregon Inlet, 63 km north of Cape Hatteras, is the only opening in the nearly 200 km between Cape Henry and Cape Hatteras that bounds the Hatteras Littoral Cell. Oregon Inlet is migrating south at an average rate of 23 m/yr. and landward at a rate of 5 m/yr. The net southerly longshore transport of sand in the vicinity of Oregon Inlet is between one-half million and one million m³/yr.

Averaged over the 160 km from False Cape to Cape Hatteras, sea-level rise accounts for 21 percent of the measured shoreline recession of 1.4 m/yr. Analysis of the budget of sediment indicates that the remaining erosion of 1.1 m/yr. is apportioned among overwash processes (31 percent), longshore transport out of the cell (17 percent), windblown sand transport (14 percent), inlet deposits (8 percent), and removal by dredging at Oregon Inlet (9 percent). This analysis indicates that the barrier system moves as a whole so that the sediment balance is relative to the moving shoreline (Lagrangian grid). Application of a continuity model to the budget suggests that, in places such as the linear shoals off False Cape, the barrier system is supplied with sand from the shelf.

Marginal sea coasts are characterized by more limited fetch and reduced wave energy. Accordingly, river deltas are more important sources of sediment than the area along the mid-Atlantic coast. Otherwise, barrier island rollover processes are quite similar. Along both coasts, offshore mining of sand may become an important source of beach nourishment.

Arctic Coasts

Tectonically, Arctic coasts are of the stable, trailing-edge type, with wide shelves backed by broad coastal plains built from fluvial and glacial deposits. Tidal amplitudes are small, and both ice and water motion are controlled predominantly by the wind. The Coriolis effect of east and west blowing winds results in water-level increases and decreases in excess of 1 meter.

At 70 degrees north latitude, the sun does not rise for seven weeks in winter and does not set for over ten weeks in summer. During the nine months of winter, the coast is frozen fast so that coastal processes are entirely cryogenic and dominated by ice-push phenomena. Wind stress and ocean currents buckle and fracture the frozen pack ice into extensive, grounded, nearshore, pressure-ridge systems known as stamukhi zones. The keels from the individual pressure ridges groove and rake the bottom, plowing sediment toward the outer barrier islands. During the three months of summer, the ice pack withdraws from the Beaufort Sea coast forming a 25-km to 50-km wide coastal waterway.

In contrast to winter, the summer processes are classical nearshore phenomena driven by waves and currents as shown by the beaches and barrier island chain in the vicinity of Prudhoe Bay (Figure 6.7). The sediment sources include river deltas, onshore ice-push across the shelf, and thaw-erosion of the low-lying permafrost seacliffs. Thaw-erosion rates of the shoreline are typically 5 to 10 m/yr. in arctic Russia and, over a 30-year period, averaged 7.5 m/yr. for a 23-km coastal segment of Alaska's Beaufort Sea coast (Reimnitz and Kempema, 1987).

The Flaxman Barrier Island chain extends westward from the delta of the Canning River. It appears to be composed of sand and gravel from the river, supplemented by ice-push sediments from the shelf (Figure 6.7). The prevailing easterly waves move sediment westward from one barrier island to the next. The channels between islands are maintained by setdown and setup currents associated with the Coriolis effect on the wind-driven coastal currents.

Coral Reef Coasts

These coasts result from the biogenic activity of the fringing reefs, which, in turn, depend on special latitudinal conditions. The configuration of the reef platforms themselves incorporates the nearshore circulation cell into a unique littoral cell (Figure 6.8). The circulation of water and sediment is onshore over the reef and through the surge channels, along the beach toward the awas (return channels), and offshore out the awas. An awa is equivalent to a rip channel on the sandy beaches of other coasts (Inman et al., 1963).

In the unique situation of coral reef coasts, the corals, foraminifera, and calcareous algae are the sources of sediment. The overall health of the reef community determines the supply of beach material. Critical growth factors are light, ambient temperature, and nutrients. Turbidity and excessive nutrients are deleterious to the primary producers of carbonate sediments. On a healthy reef, grazing reef fishes bioerode the coral and calcareous algae and contribute sand to the transport pathway onto the beach.

The beach acts as a capacitor storing sediment transported onshore by the reef-moderated wave climate. It buffers the shoreline from storm waves, and releases sediment to the awas. In turn, the awas channel runoff and turbidity away from the reef flats and out into deep water. Where the reef is damaged by excessive terrigenous runoff, waste disposal, or overfishing, the beaches are imperiled.

81

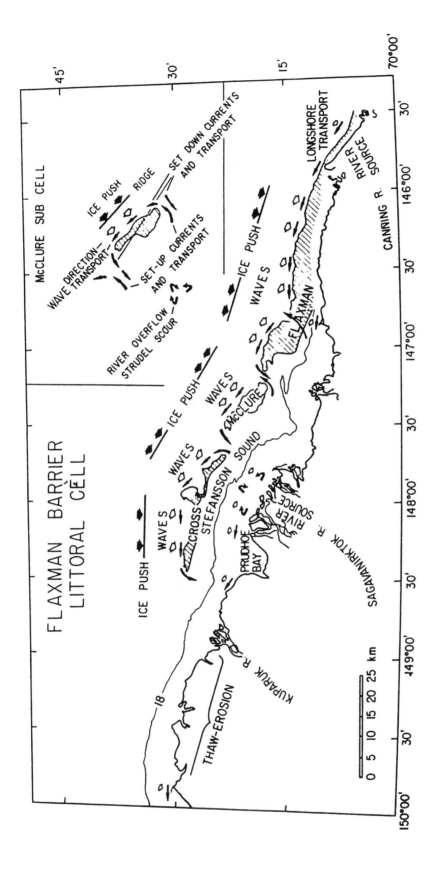

FIGURE 6.7 Flaxman Barrier Littoral Cell extending from Brownlow Point off the Canning River to Reindeer Island, Alaska. Sediment sources, transport paths and mechanisms are shown schematically. Hachured areas defined by the 18-foot depth contour.

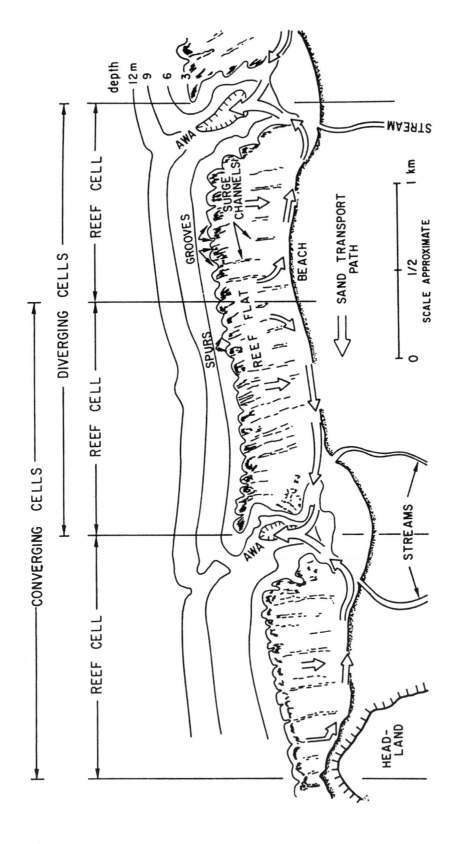

FIGURE 6.8 Schematic diagram of littoral cells along a fringing reef coast.

REFERENCES

Bard, E. 1990. Calibration of the ^{14}C timescale over the past 30,000 years using mass spectrometric U-Th ages from Barbados corals. Nature 345:405.

Barnett, T. P. 1984. The estimation of "global" sea level changes: a problem of uniqueness. Journal of Geophysical Research 89(C5):7980-88.

Barry, R. G. 1981. Trends in snow and ice research. Elsevier Oceanography Series, 62(46): 1139-44.

Curray, J. R. 1960. Sediments and History of Holocene Transgression, Continental Shelf, Northwest Gulf of Mexico. Pp. 221-266 in F. P. Shepard, F. B. Phleger, and Tj. H. van Angel, eds, Recent Sediments, Northwest Gulf of Mexico, 1951-1958. Tulsa, Oklahoma: American Association of Petroleum Geologists.

Curray, J. R. 1961. Late Quaternary sea level: a discussion. Geological Society of America Bulletin 72:1707-12.

Curray, J. R. 1965. Late Quaternary History; Continental Shelves of the United States, Pp. 623-35 in H.E. Wright, Jr. and D.G. Frey, eds., The Quaternary of the United States. Princeton, New Jersey: Princeton University Press.

Emery, K. O. 1980. Relative sea levels from tide-gauge records. National Academy of Sciences, Proceedings 77(12):6968-72.

Fairbanks, R. G. 1989. A 17,000-year glacio-eustatic sea level record. Nature 342:637-642.

Inman, D. L., and B. M. Brush. 1973. The coastal challenge. Science, 181:20-32.

Inman, D. L., and R. Dolan. 1989. The Outer Banks of North Carolina: Budget of sediment and inlet dynamics along a migrating barrier system. Journal of Coastal Research, 5(2):192-237.

Inman, D. L., and P. M. Masters. 1991. Budget of Sediment and Prediction of the Future State of the Coast. Chapter 9 in State of the Coast Report, San Diego Region Coast of California Storm and Tidal Waves Study, v.1. Final Report, September 1991. U.S. Army Corps of Engineers, Los Angeles District. Chapters 1-10, v. 2, Appen, A-I.

Inman, D. L., and C. E. Nordstrom. 1971. On the tectonic and morphologic classification of coasts, Journal of Geology 79(1):1-21.

Inman, D. L., W. R. Gayman, and D. C. Cox. 1963. Littoral sedimentary processes on Kauai, a sub-tropical high island. Pacific Science 17(1):106-130.

Inman, D. L., R. J. Tait, and C. E. Nordstrom. 1971. Mixing in the surf zone. Journal of Geophysical Research 76(15):3493-3514.

Inman, D. L., C. E. Nordstrom, and R. E. Flick. 1976. Currents in Submarine Canyons: An Air-Sea-Land Interaction. Pp. 275-310 in M. Van Dyke et al., eds., Annual Review of Fluid Mechanics, v. 8.

Leatherman, S. P., ed. 1979. Barrier Islands from the Gulf of St. Lawrence to the Gulf of Mexico. New York, New York: Academic Press.

Leatherman, S. P., ed. 1981. Overwash Processes. Benchmark Papers in Geology, v. 58. Stroudsburg, Pennsylvania: Hutchinson Ross Publishing Company.

National Research Council. 1987. Responding to Changes in Sea Level: Engineering Implications. Marine Board, National Research Council. Washington, D.C.:National Academy Press.

Peltier, W. R. 1986. Deglaciation-induced Vertical Motion of the North American Continent and Transient Lower Mantle Rheology. Journal of Geophysical Research 91(B9):9099-9123.

Reimnitz, E., and E. W. Kempema. 1987. Thirty-Four-Year Shoreface Evolution at a Rapidly Retreating Arctic Coastline. Pp. 161-164 in T. D. Hamilton and J. P. Galloway eds. Geologic Studies in Alaska by the U. S. Geological Survey During 1986: United States Geological Survey Circular 998.

7

Landscapes and the Coastal Zone

R. Eugene Turner
Louisiana State University
Baton Rouge, Louisiana

Each shade of blue or green sums up in itself a structure and a history, for each lake is a small world, making its nature known to the larger world of the desert most clearly in its colour. These little worlds of turquoise, set among red, brown, grey and white rocks, are not independent of the dry landscape around them.... In the quality of this scene, accentuated by the foetid sulphurous water that lies at the bottom of the lake, may be traced the whole life of the surrounding country (The Clear Mirror, G. E. Hutchinson, 1936).

INTRODUCTION

Scientists and resource managers often look at the coastal zone without referring to the larger landscape in a quantitative way. A conspicuous management practice illustrating the dichotomy between the landscape scale of examination and management action involves permit decisions. The typical Federal Section 404 permit is evaluated and issued one permit at a time, without serious consideration of the cumulative effects of many permit decisions, thus thwarting the apparent intent of several legal instruments of the federal and state resource management (e.g., Bedford and Preston, 1988; Gosselink et al., 1990). The influence of landscapes and landscape-scale changes may originate either in situ or far from the immediate area of concern and often in unappreciated ways. The theory of island biogeography (MacArthur and Wilson, 1967) is one example of a fruitful quantitative analysis of how the size, relationship, and shape of landscape patches affect species distribution. Process-oriented landscape studies are, in contrast, rare. It is my purpose here to illustrate various landscape interactions of coastal ecosystems and thereby encourage integrated analysis and management.

ESTUARINE VARIABILITY IN THE UNITED STATES

The geomorphology of U.S. estuaries varies enormously, and this variability influences estuarine functions. Examples of estuarine variability are found in Figure 7.1. The drainage-basin size, water-surface size, freshwater turnover time, and water–marsh ratio varies by several orders of magnitude from the Maine estuaries around the coast to the state of Washington. Estuarine functions are strongly influenced by this variability, which is, of course, overlain with seasonal, annual, and daily cycles. For example, as estuarine flushing time decreases, the likelihood of water pollution problems increases. Fisheries are limited by wetland area because of requirements for refuge or food (e.g., Turner, 1992; Turner and Boesch, 1987; Figure 7.2). We know from freshwater studies that landscape pattern may influence bird distributions (Figure 7.3; Brown and Dinsmore, 1986). Examples of area–species relationships for coastal communities are less frequently documented than for terrestrial ecosystems, perhaps because of the variability of estuarine flushing. This variability makes data collection and analysis more difficult. However, some of the earliest studies of species colonization and turnover were done on mangrove islands (by D. Simberloff and colleagues), and the principles should be generally applicable to the coastal zone. For example, habitat diversity does appear to be related to habitat area (Figure 7.4).

One cannot easily clone landscapes as in the usual scientific experiment. However, when accomplished, the results may be dramatically illustrative (e.g. Schindler, 1977; Likens, 1992). Comparisons between landscape units are often fruitful, and modelling has a role. Some the clearest cases of landscape-scale influences come from documenting recent changes and the cause-and-effect conclusions based on strong inferences. Three case studies of landscape changes are briefly discussed below. In one case (Everglades), the influence of variable water-flow patterns are discussed. The second case (Mississippi River) illustrates the probable role of changing riverine nutrient flows and consequences to the continental shelf ecosystem. The third example (coastal Louisiana) discusses the indirect impacts of many small dredge and fill permit decisions—an *in situ* change—on wetland losses.

CASE HISTORY: SOUTH FLORIDA HYDROLOGIC CHANGES

The ecosystems of south Florida are very sensitive to the fluctuations and duration of soil moisture, flooding, and drying cycles. They are also nutrient-poor systems, especially for phosphorus (e.g., Ornes and Steward, 1973). The natural water balance, in turn, is driven by seasonal and long-term rainfall, evaporation, plant transpiration, below ground seepage, land elevation, soil infiltration, and natural channel flow. Water movement from Lake Okeechobee (approximately +5 m msl) southward for 70 miles toward the Miami area (approximately +2 m msl) is sensitive to the elevation gradients. Water movement eastward was formerly only through the coastal ridges at high water stages. The westward movement is impeded by sandy ridges averaging about +8 m msl on the western border of the Everglades Agricultural Area (south of Lake Okeechobee). High water is a flood problem for the 3 million residents near Miami, and low water is a threat to the water table and freshwater supplies.

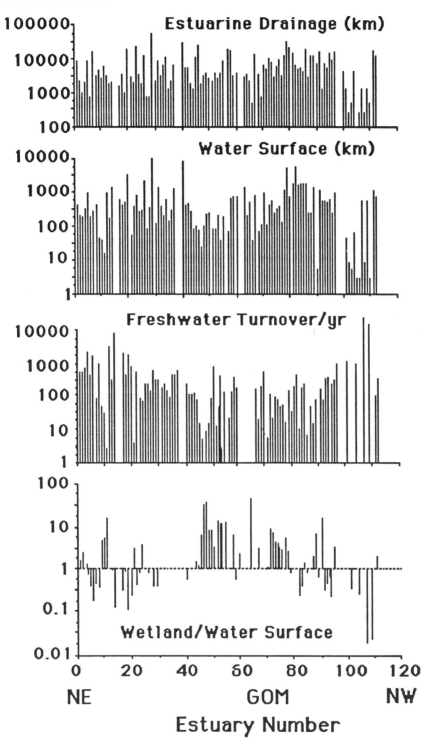

FIGURE 7.1 Variations in the morphology of U.S. estuaries. NE is northeast United States; GOM, Gulf of Mexico; NW, northwest United States.

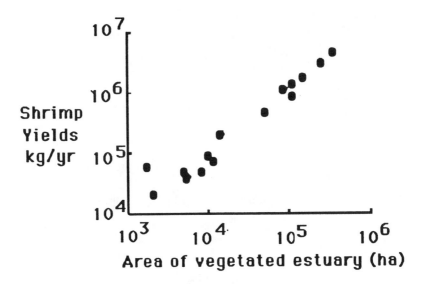

FIGURE 7.2 The relationship between intertidal vegetation and penaeid shrimp yields from the estuaries of the northern Gulf of Mexico. (Source: Turner and Rabalais, 1992).

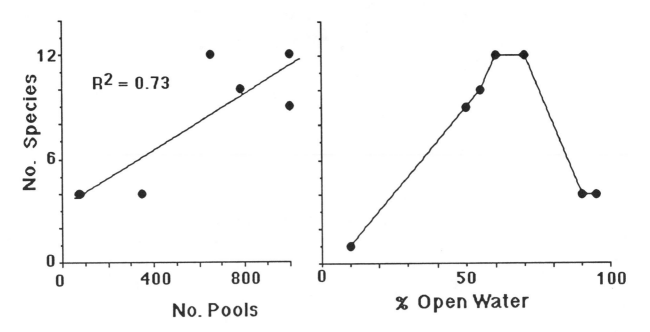

FIGURE 7.3 Species richness of birds in freshwater ponds in relationship to cover-water ratios expressed as (1) the number of pools in the emergent marsh and (2) the percent open water. (Source: Weller and Fredrickson, 1974). Reprinted with persmission from Cornell University, Laboratory of Orninthology, 1974.

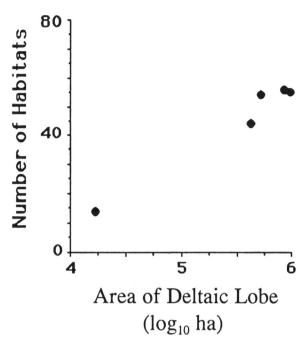

FIGURE 7.4 Habitat number vs. size of the Mississippi River deltaic coastal plain. (Source: Neill and Deegan, 1968) Reprinted with permission from American Midland Naturalist, 1968.

Man has been altering this ecosystem for most of this century. The history of Florida's development (see Table 7.1), particularly south Florida, began with a simple interest to drain the excess water and then to occupy the land for agriculture. It soon became evident that the limitations of excess water were also matched by the problems of water scarcity. What at first was a plan to drain the swamp by the state became a complex water-management scheme involving local, regional, state, and national agencies. A diversity of perspectives, priorities, and monies accompanies this complexity. It is a difficult area to manage because it includes a National Park, which is a conservation zone with site-specific and exclusive land use regulations. Human enterprises are inherently interested in stable water supplies, whereas the Everglades ecosystem is very responsive and requires fluctuating water supplies. These land use changes have led to the loss or destruction of about one-half of the natural Everglades ecosystem through either direct habitat loss or the indirect influences of development. In particular, water movement through and within the naturally defined ecosystem has been modified through land use, land drainage, and both water diversions and storage (e.g., in 1990 there were more than 1400 miles of canals, 65 major spillways, and hundreds of water control structures; Figure 7.5). Compared with the 1900 conditions, plant coverage is vastly reduced or changed (e.g., Steward and Ornes, 1973a, b; Hagenbuck et al., 1974), local weather patterns may have been altered, [and there is increased coastal saltwater intrusion and fire;] animal populations dependent on plants and their predators/decomposers have been put at risk (Tables 7.1 and 7.2).

Some specific effects of the altered hydrology are

TABLE 7.1. Drainage Issues in South Florida (Based mostly on Davis, 1943, Parker, et al., 1955; Anon, 1948; Stephens, 1969; Tebeau, 1971

---

pre 1920s	unregulated water level in Lake Okeechobee. Lake size larger than present, filled more of its basin, and wider fluctuations in water level than presently. Water did not overflow until >20.5 msl, and over the southern rim. The Everglades were already wet before the overflow in late summer or early fall.
1940	peat fires more frequent, soil oxidation even to bare rock is obvious, concern about relationship of 1935 and 1939 freeze and how air temperature is related to water level.
1950	the St. Lucie Canal and Caloosahatchee Canal and River system are artificial outlets diverting water from Lake Okeechobee to the Atlantic and Gulf of Mexico, respectively. Rainwater is major source of water for Everglades. Saltwater intrusion into the freshwater aquifer below msl is evident (e.g., Dade County).
1970s	coastal ground-water levels significantly reduced from the 1940s (e.g., Klein 1973); Big Cypress waterflow changed from slow, prolonged southward sheetflow overland to accelerated and shortened-period runoff through canal system; water quality problems (e.g., algal blooms) in Lake Okeechobee. The 'rainfall' plan went into effect guaranteeing water to the Everglades on a regulated basis.
1990	more than 1,400 miles of canals. Cattails found on 6,000 acres of Loxahatchee Wildlife Refuge (WCA-1) and another 24,000 acres show elevated phosphorous levels; various restoration efforts underway, including the restoration of the Kissimmee River Basin; annual flow distribution in Shark Slough (a principle water entry point into the Everglades National Park) changed from a mixed east/west pattern, to almost exclusively western slough (Johnson and Ogden, 1990). Lake Okeechobee water levels now 12.6 to 15.6 msl, compared to +18 to 20 msl at the turn of the century; flow over the southern rim is vastly restricted (based mostly on Davis, 1943; Parker et al., 1955; Anon, 1948; Stephens, 1969; and Tebeau, 1971).

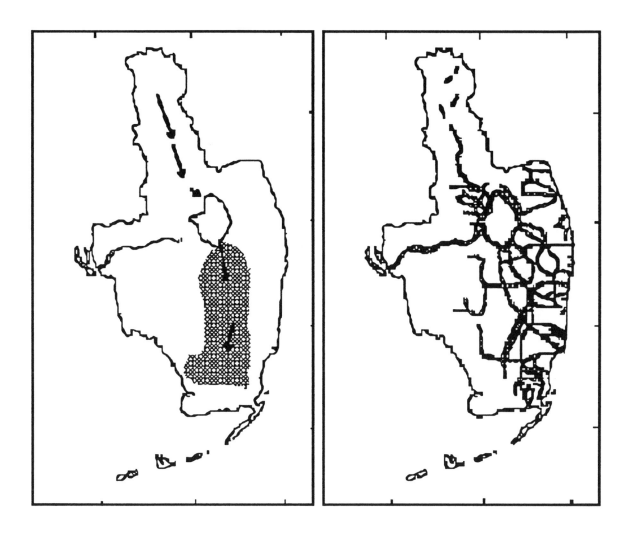

FIGURE 7.5 The hydrology of south Florida around 1900 (left) and presently (right).

TABLE 7.2   Some consequences of recent hydrologic changes to south Florida coastal communities.

Species	Natural Conditions	Recent	Notes
Area:	100%	36% urban/agriculture	
		32% impounded	
		12% overdrained	
• Hydrology			
• hydroperiod:	gradual	shorter	
• water height:	seasonal	higher in impoundments	
		seasonality reduced	
• Atlantic diversions:	minor	major	
Alligator			
• population size:	178,000	116,300 (1971-1982)	M. Fleming, unpublished based on computer model
• percent occurrence of nest flooding:	7%	19%	
Wading birds, nesting in central-south Florida:	77-93,000 (1931-40)	9,850 (1980-89)	Ogden and Johnson, 1992
Wood stork			
• nesting success:	78% (1953-61)	21% (1962-1989)	Ogden and Johnson, 1992
• colony formation in Nov/Dec:	87% (1953-69)	10% (1970-89)	Ogden and Johnson, 1992

- exotic species are now in the Everglades National Park (e.g., Australian pine [Melaleuca sp.], Brazilian pepper, hydrilla, water hyacinth, cajeput, walking catfish, pike killifish, and several species of cichlides);
- soil subsidence and water-level declines since 1900 through much of the ecosystem are measured in feet, not inches (Alexander and Crook, 1973);
- agricultural fertilizers leak to downstream ecosystems;
- fires are more frequent (e.g., Stephens, 1969);
- perhaps a dozen species whose recovery is compromised by the altered hydrologic cycles are endangered or threatened by extinction;
- significant legal issues arise over the allocation of water resources and the quality of that water;
- the long-term sustainability of drained agricultural lands is in doubt;
- saltwater intrusion is a threat to urban water supplies; and,
- changes in animal populations occur for those species dependent on plants and their predators/decomposers.

CASE HISTORY: MISSISSIPPI RIVER WATERSHED NUTRIENT ADDITIONS

The Mississippi River watershed is 41 percent of the area of the contiguous 48 states (Figure 7.6). In terms of length, discharge, and sediment yield, the main river channel is, respectively, the third, eighth, and sixth, largest river in the world (Milliman and Meade, 1983). The river has been shortened by 229 km in an effort to improve navigation, and it has a flood control system of earthwork levees, revetments, weirs, and dredged channels for much of its length that has isolated most riverine wetlands from the main channel and left them drier.

Changes in three indicators of water quality were documented by Turner and Rabalais (1991) and are presented here: phosphorus (as total phosphorus), silicon (as silicate), and dissolved inorganic nitrogen (as nitrate). All three are important nutrients for freshwater and marine phytoplankton growth and production. The pervasive relationships between phosphorus and freshwater phytoplankton communities is well established (e.g., Schindler, 1977, 1988; Vollenweider and Kerekes, 1980). Diatoms, an important food species for freshwater and marine fish and invertebrates, require silicon to build their nests. Schelske et al. (1983, 1986) proposed that increased phosphorus loading in lakes stimulated diatom production with the subsequent loss of silicon (as diatom tests) when deposited in sediments. Eventually a new steady-state silicate concentration develops in the water column where diatoms are less numerous and their growth is silicon limited. Nitrogen often acts in concert with phosphorus to regulate phytoplankton communities in freshwater ecosystems and may often be the dominant nutrient limiting phytoplankton of estuarine and marine communities (e.g., Valiela, 1984; D'Elia et al., 1986; Harris, 1986).

The mean annual concentration of nitrate in the lower Mississippi River was about the same from 1905 to 1906 and from 1933-1934 as in the 1950s but has subsequently doubled (Figure 7.7). The mean annual concentration of silicate was about the same from 1905 to 1906 as in the 1950s and then

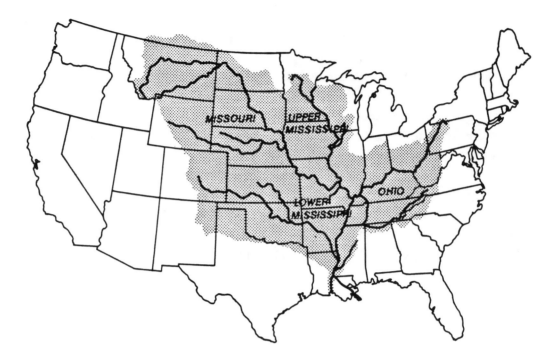

FIGURE 7.6 The drainage basin of the Mississippi River.

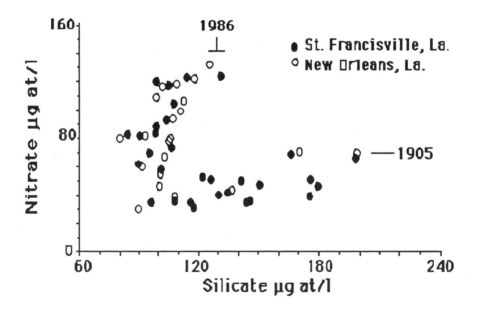

FIGURE 7.7 Average annual nitrate and silicate concentration in the lower Mississippi River. The data are from the U.S. Geological Survey and the New Orleans Water Board. (Source: Turner and Rabalais, 1991). Reprinted with permission from American Institute of Biological Sciences, 1991.

declined by 50 percent. The concentration of silicate increased from 1985 to 1988, whereas the concentration of nitrate decreased slightly in the same period. Although the concentration of total phosphorus appears to have increased since 1972 (the earliest records that could be found), the variations between years are large, and the trends, if they exist, are not clear.

The silicate/nitrate atomic ratios in the lower Mississippi River for this century have changed as the concentrations varied. The silicate/nitrate atomic ratio was about 1:4 at the beginning of this century, 1:3 in 1950, and then rose to about 1:4.5 over the next 10 years before plummeting to around 1:1 in the 1980s.

The seasonal patterns in nitrate and silicate concentration also changed. There was no pronounced peak in nitrate concentration earlier this century, whereas there was a spring peak from 1975 to 1985 (Figure 7.8). A seasonal peak in silicate concentration, in contrast, is no longer evident. There was no marked seasonal variation in total phosphorus from 1975 to 1985.

The likely causal agent of these changes is the widescale and intensive use of nitrogen and phosphorus fertilizers, which reached a plateau in the 1980s. The current consumption of phosphorus fertilizer use in the United States is stable but is increasing throughout the world (Figure 7.9). There is a direct relationship between annual nitrogen fertilizer use and nitrate concentration in the river (Figure 7.10). An indirect relationship between phosphorus fertilizer use and silica concentration in the river is also observed, as predicted by the hypotheses of Schelske et al. (1983, 1986).

The combination of changes in nitrate, phosphorus and silicate has almost certainly influenced the coastal marine phytoplankton community (in particular, leading to a decline in diatom abundance), and has perhaps led to increased phytoplankton production. This is especially likely if a community is nitrogen limited, as many coastal systems are thought to be. It is not clear, however, if larger or more severe hypoxic zones have formed in bottom waters offshore (Rabalais et al., 1991) as a result of these riverine water-quality changes.

These changes are important to understand, if only because nitrogen is commonly thought to be limiting phytoplankton growth in coastal and oceanic waters (e.g., Harris, 1986; Valiela, 1984). The abundance of coastal diatoms is influenced by the silicon supplies, whose silicon/nitrogen atomic ratio is about 1:1. Diatoms out-compete other algae in a stable and illuminated water column of favorable silicate concentration. When nitrogen increases and silicate decreases, flagellates may increase in abundance (Officer and Ryther, 1980) and form undesirable algal blooms. In particular, noxious blooms of flagellates are becoming increasingly common in coastal systems. Zooplankton, important diatom consumers and a staple of juvenile fish, are thus affected by these nutrient changes in a cascading series of interactions. Furthermore, where eutrophication occurs, hypoxia often follows, presumably as a consequence of increased organic loading. Supportive evidence of this benthic-pelagic coupling is found in the observations of Cederwall and Elmgren (1980), who demonstrated a rise in macrobenthos around the Baltic islands of Gotland and Oland, which they attributed to eutrophication, a known event (Nehring, 1984).

However, not all coastal systems are nitrogen limited (e.g., the Huanghe in China is phosphorus limited; Turner et al., 1990), nor is changing nutrient loading the only factor influencing phytoplankton growth (Skreslet, 1986). Marine phytoplankton may also respond in various ways to nutrient additions introduced gradually or suddenly, with changing flushing rates or salinity, and with

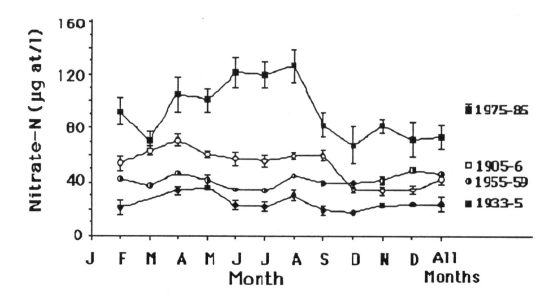

FIGURE 7.8 Seasonal nitrate concentrations in the lower Mississippi River. A $^+/_-$ 1 standard error of the mean is shown. Source: Turner and Rabalais, 1991). Reprinted with permission from American Institute of Biological Sciences, 1991.

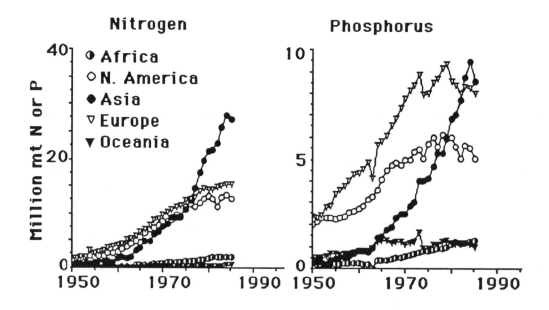

FIGURE 7.9 World nitrogen and phosphorus fertilizer consumption since 1950. (Source: Turner and Rabalais, 1991). Reprinted with permission from American Institute of Biological Sciences, 1991.

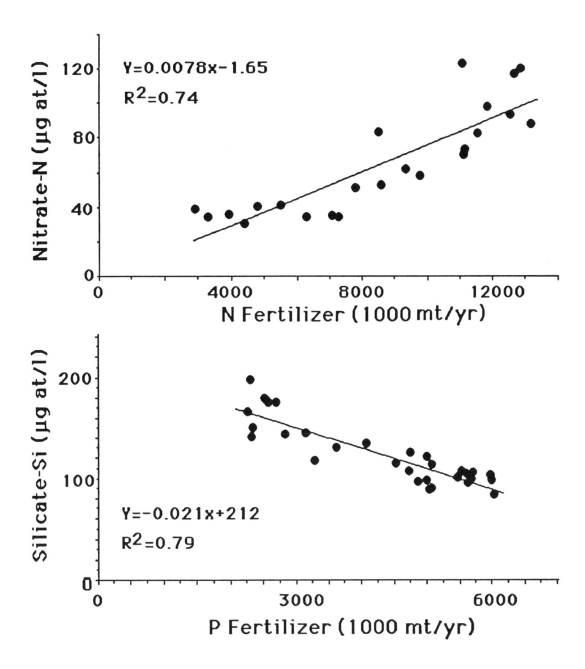

FIGURE 7.10 The relationship between fertilizer use and water quality at St. Francisville, Louisiana. Top: Nitrogen (as N) fertilizer use in the United States and average annual nitrate concentration from 1960 to 1985. Bottom: phosphorus (as P_2O_5) fertilizer use in the United States and average silicate concentration at St. Francisville, Louisiana from 1950 to 1987. (Source: Turner and Rabalais, 1991). Reprinted with permission from American Institute of Biological Sciences, 1991.

cell density (Sakshaug et al., 1983; Sommer, 1985; Suttle and Harrison, 1986; Turpin and Harrison, 1980).

Management of eutrophication on a national scale has not sufficiently integrated freshwater and estuarine systems. The national freshwater policy is to control phosphorus, and it is based on the numerous excellent laboratory and field studies of the stimulatory effect of phosphorus on freshwater ecosystems. However, the analysis and management of nitrogen limited coastal systems is becoming more complicated as the eutrophication modifies the transition zone between phosphorus and nitrogen-limited aquatic ecosystems. A national policy common to both freshwater and coastal systems is sewerage treatment. But, as is shown for the Mississippi River (Turner and Rabalais, 1991), the terrestrial system is very leaky, and treatment does not mean a reduction of loading to the estuary via water and precipitation. A second understated issue, therefore, is that sewerage treatment upstream does necessarily equate to controlling nutrient loading to downstream estuaries.

A third point is that mitigation of nutrient applications seems a less prudent management policy, compared with an outright reduction in use. The ecosystem is simply too leaky to control all nutrient flows between the application site and estuary.

CASE HISTORY: LOUISIANA COASTAL WETLAND DREDGE AND FILL

Dredging is a conspicuous human activity affecting Louisiana's coastal wetlands. It is principally related to oil and gas recovery efforts and results in large areas of canals and residual spoil deposits, or *spoil banks* (80,426 ha, equivalent to 6.8 percent of the wetland area in 1978; Turner, 1990; Baumann and Turner, 1990; Figure 7.11). The aggregate length of these spoil banks in Louisiana is in the neighborhood of 12,000 miles. To remove all of them would cost about as much as to build three river diversions, that is, about $500 million.

There are strong and probably partially reversible cause-and-effect relationships between wetland losses and these hydrologic changes. Canals and spoil banks are the most likely cause of at least 30–59 percent of Louisiana's coastal wetland losses from 1955 to 1978 (51,582 ha/yr, or 0.85 percent/yr; Turner and Cahoon, 1987). Wetland losses may be due to either direct or indirect impacts of spoil banks and canals. Sixteen percent of these wetland losses resulted from the impacts of dredging wetlands into open water and spoil bank; at least 14–43 percent of these wetland losses were the result of the indirect impacts of spoil banks and canals on tidal water movement into and out of the wetlands (Swenson and Turner, 1987). About 13 percent of the direct wetland losses were due to agricultural and urban expansion into wetlands. Indirect impacts result from the longer wetland drying cycles, even in semi-impounded wetlands, as a consequence of altered water movements into and out of the wetland. The lengthened drying periods promote soil oxidation and subsequent soil shrinkage (Table 7.3). Flooding events may also lengthen behind spoil banks, presumably as a consequence of water being trapped behind the spoil bank once water enters overland during very high tides. The impacts also result from lower sedimentation rates behind spoil banks in any wetland type, probably because of the reduced frequency and depth of tidal inundation (Figure 7.12), and increased waterlogging of soils that changes soil chemistry. Plants may become stressed to the point where

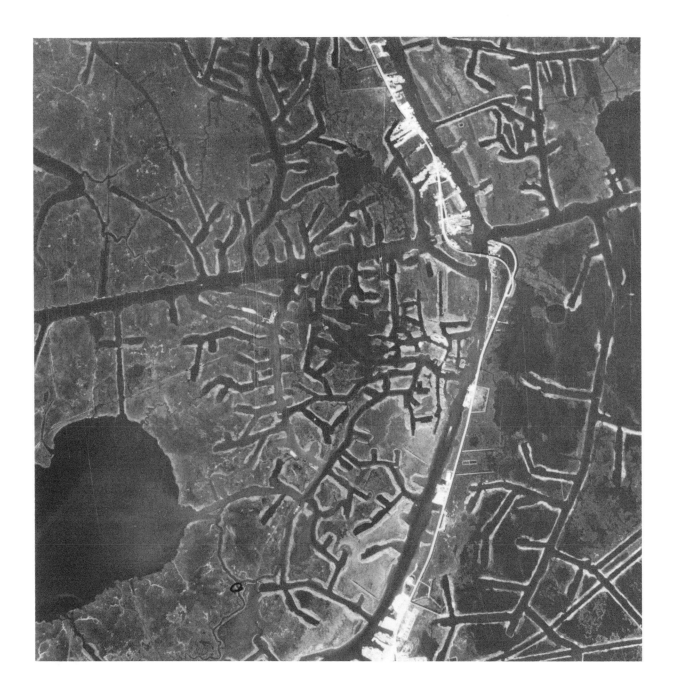

FIGURE 7.11 Aerial photograph of Louisiana coastal wetland. The dredged canals are usually straight, and the spoil is piled alongside in a continuous line.

TABLE 7.3 Changes in hydrologic regime of a semi-impounded saltmarsh

	Control	Semi-Impounded
Flooding		
number events per month	12.9	4.5
event length (hours)	29.7	149.9
Drying		
number events per month	11.6	4.00
event length (hours)	31.2	53.9
Mean Water Level (cm; annual average)	1.71	3.99
Volume Exchange (wetland surface)		
above ground	0.15	0.06
below ground	0.09	0.04

Source: Swenson and Turner, 1987

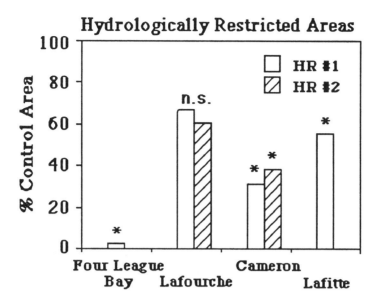

FIGURE 7.12 Vertical accretion rates in two hydrologically restricted areas (HR #1 and HR #2) compared with control sites nearby. Data were normalized to the control site values (100%). An "*" by the bar indicates a statistically significant difference between the hydrologically restricted site and the control site. (Source: Cahoon and Turner, 1989)

growth reduction or even die-back occurs (e.g., Babcock, 1967; King et al., 1982; Wiegert et al., 1983; Mendelssohn and McKee, 1987). In addition, the spoil banks consolidate the underlying soils. Water movements below ground are thus decreased, both because of the reduced cross-sectional area and the reduced permeability of material beneath the levee.

The combined effects of sediment deprivation, increased wetland drying, and lengthened soil flooding result in a hostile soil environment for plants. The death of plants reduces sediment trapping amongst the plant stems and accumulation of plant material at the soil surface and below ground. Small, shallow ponds may form and enlarge due to scouring under even light winds. The practical consequence of these causal mechanisms is a strong, direct relationship between wetland losses and canal density on a local and coastwide basis (e.g., Turner and Rao, 1990; Figure 7.13).

SUMMARY

Landscapes are a functioning legacy of many broad influences, including geology, climate, biological evolution, etc. Landscapes are more than reactive parts that can be understood and managed in isolation from each other. Many relationships such as edge configuration, habitat area, fragmentation, and human use are reflected in the ecological functions of the parts and of the whole.

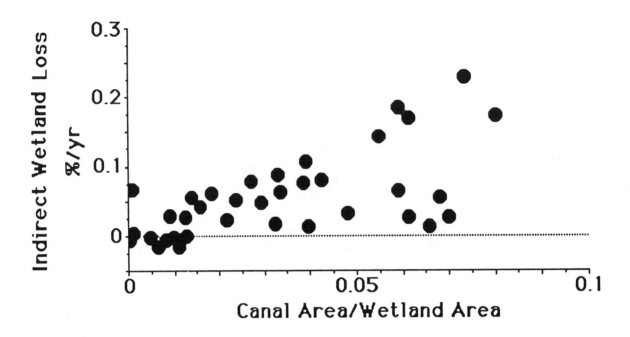

FIGURE 7.13 The relationship between canal density and indirect wetland loss rates (wetland change excluding the area lost to canals and development). Area calculations are for changes within 7 1/2 ft. quadrangle maps from 1955 to 1978 for the St. Bernard, Barataria Bay, and Terrebonne Bay estuaries. Only areas with more than 25 percent wetland in the quadrangle maps are included. Total area included equals 413,000 ha.

We must learn to live with the fact that humans are changing landscapes in a multitude of ways. The details may appear as a painting by Cezanne viewed with a magnifying glass--the colorful tones and textures of oil or pastels. But the painting quality, the forms, and the impressions are intended to be considered from further away. Both the natural and managed coastal-zone parts must be viewed in this context. Individual permits and individual species are like the small brushstrokes. They are required for the whole picture, but the picture quality does not emerge completely unless there is a broad and fully encompassing view. Coastal zones must be viewed not only from close up but from a distance. Most management excludes that landscape view. The consequences of many small decisions tend to be overwhelming, consistent with the acceptance of the fallacy of multiple uses. Landscape fragmentation and loss of functions will result in the absence of a broad view. This is the history of landscapes (e.g., Hoskins, 1970). However, as with a Cezanne painting, the brushstrokes can be changed and managed, one by one, to result in a more masterly painting. This will be most likely accomplished with less arrogance about our abilities to substitute intense manipulation for the multitude of natural interactions.

REFERENCES

Alexander, T. R., and A. G. Crook. 1973. Recent and Long-Term Vegetation Changes and Patterns in South Florida. U. S. National Park Service PB-231 939.

Anon. 1948. Soils, geology, and water control in the Everglades Region. Florida Agricultural Experiment Stations Bulletin.

Babcock, K. M. 1967. The Influence of Water Depth and Salinity on Wiregrass and Salt-marsh Grass. Unpubl. Ph.D. Diss. School Forestry and Wildlife Management, Baton Rouge, Louisiana: Louisiana State University.

Baumann, R. H., and R. E. Turner. 1990. Direct impacts of outer continental shelf activities on wetland loss in the central Gulf of Mexico. Environmental Geology and Water Resources 15:189-198.

Bedford, B. L. and E. M. Preston, eds. 1988. Cumulative effects on landscape systems of wetlands. Environmental Management 12(5).

Brown and Dinsmore. 1986. Implications of marsh size and isolation for marsh bird management Journal of Wildlife Management 50:392-397.

Cahoon, D. R., and R. E. Turner. 1989. Accretion and canal impacts in a rapidly subsiding wetland. 11. Feldspar marker horizon technique. Estuaries 12:260-268.

Cederwall, H., and R. Elmgren. 1980. Biomass increases of benthic macrofauna demonstrates eutrophication of the Baltic Sea. Ophelia, Supplement 1:287-304.

Davis, J. H., Jr. 1943. The Natural Features of Southern Florida; Especially the Vegetation, and the Everglades. Florida Geological Survey Geological Bulletin 25. Tallahassee.

D'Elia, C. J., J. G. Sanders, and W. R. Boyton. 1986. Nutrient enrichment studies in a coastal plain estuary: Phytoplankton growth in large-scale, continuous cultures. Canada Journal Fisheries Aquatic Science 43:397-406.

Gosselink, J. G., G. P. Shaffer, L. C. Lee, D. M. Burdick, D. L. Childers, N. C. Leibowitz, S. C. Hamilton, R. Boumans, D. Cushman, S. Fields, M. Koch, and J. M. Visser. 1990. Landscape conservation in a forested wetland watershed. Bioscience 40:588-600.

Hagenbuck, W. W., R. Thompson, and D. P. Rodgers. 1974. A Preliminary Investigation of the Effects of Water Levels on Vegetative Communities of Loxahatchee National Wildlife Refuge, Florida. U.S. Bureau Sport Fisheries and Wildlife. PB-231 61 1.

Harris, G. P. 1986. Phytoplankton Ecology: Structure, Function and Fluctuations. New York, New York: Chapman and Hall.

Hoskins, W. G. 1970. The Making of the English Landscape. London: Pelican Books.

Johnson, R. A. and J. C. Ogden. 1990. An Assessment of Hydrological Improvements and Wildlife Benefits from Proposed Alternatives for the U.S. Army Corps of Engineers' General Design Memorandum for Modified Water Deliveries to Everglades National Park. U.S. Park Service, unnumbered document.

King, G. M., M. J. Klug, R. G. Wiegert, and A. G. Chalmers. 1982. Relation of soil water movement and sulfide concentration to Spartina alterniflora production in a Georgia salt marsh. Science 218:61-63.

Klein, H., J. T. Armbruster, B. F. McPerson, and H. J. Freiberger. 1973. Water and the South Florida Environment. U. S. Geological Survey PB-236 951.

Likens, G. E. 1992. The Ecosystem Approach: Its Use and Abuse. Vol. 3, Excellence in Ecology. Oldendor/Luhe, Germany. Ecology Institute.

MacArthur, R. H., and E. 0. Wilson. 1967. The Theory of Island Biogeography. Princeton, New Jersey: Princeton University Press.

Mendelssohn, A., and K. L. McKee. 1987. *Spartina alterniflora* die-back in Louisiana: time-course investigations of soil waterlogging effects. Journal of Ecology 76:509-521.

Milliman, J. D., and R. Meade. 1983. World-wide delivery of river sediment to the ocean. Journal of Geology 91:1-21.

Nehring, D. 1984. The further development of the nutrient situation in the Baltic proper. Ophelia, Supplement 3:167-179.

Neill, C., and L. A. Deegan. 1986. The effect of Mississippi River delta lobe development on the habitat composition and diversity of Louisiana coastal wetlands. American Midland Naturalist 116:296-303.

Officer, C. B., and J. H. Ryther. 1980. The possible importance of silicon in marine eutrophication. Marine Ecology Progress Series 3:83-91.

Ogden, J. C. and R. A. Johnson. 1992. Ecosystem restoration in Everglades National Park: A Prerequisite for Wildlife Recovery. Unpublished Document.

Ornes, W. H., and K. K. Steward. 1973. Effect of Phosphorus and Potassium on Phytoplankton Populations in Field Enclosures. U.S. National Park Service PB-231 650.

Parker, G. C., G. E. Ferguson, S. K. Love, and others. 1955. Water Resources of Southeastern Florida. U. S. Geological Survey Water-Supply Paper 1255.

Rabalais, N. N., R. E. Turner, W. J. Wiseman, Jr., and D. F. Boesch. 1991. A brief summary of hypoxia on the nonhem Gulf of Mexico continental shelf: 1985-1988. Journal Geological Society of London Sp. Publ. 58:35-47.

Sakshaug, E., K. Andresen, S. Mkklestad, and Y. Olsen. 1983. Nutrient status of phytoplankton communities in Norwegian waters (marine, brackish, and fresh) as revealed by their chemical composition. J. Plankton Res. 5:175-196.

Schelske, C. L., E. F. Stoermer, D. J. Conley, J. A. Robbins, and R. M. Glover. 1983. Early eutrophication in the lower Great Lakes: New evidence from biogenic silica in sediments. Science 222:320-322.

Schelske, C. L., D. J. Conley, E. F. Stoermer, T. L. Newberry, and C. D. Campbell. 1986. Biogenic silica and phosphorus accumulation in sediments as indices of eutrophication in the Laurentian Great Lakes. Hydrobiologia 143:79-86.

Schindler, D. W. 1977. Evolution of phosphorus limitation in lakes. Science 195:260-262.

Schindler, D. W. 1988. Detecting ecosystem response to anthropogenic stress. Canadian Journal of Fisheries and Aquatic Science 44 (suppl.):435-443.

Skreslet, S., ed. 1986. The Role of Freshwater Outflow in Coastal Marine Ecosystems. New York, New York: Springer-Verlag.

Sommer, U. 1985. Comparison between steady state and non-steady state competition: Experiments with natural phytoplankton. Limnology Oceanography 30:335-346.

Stephens, J. C. 1969. Peat and muck drainage problems. American Society Civil Engineers Proc. J. Iff. Drainage Div. 95:285-305.

Steward, K. K., and W. H. Ornes. 1973a. Assessing the Capability of the Everglades Marsh Environment for Renovating Wastewater. U.S. National Park Service PB 231 652.

Steward, K. K., and W. H. Omes. 1973b. Investigations into the Mineral Nutrition of Sawgrass Using Experimental Culture Techniques. U.S. National Park Service PB 231 609.

Suttle, C. A., and P. J. Harrison. 1986. Phosphate uptake rates of phytoplankton assemblages grown at different dilution rates in semicontinuous culture. Canada Journal Fisheries Aquatic Science 43:1474-1481.

Swenson, E. M., and R. E. Turner. 1987. Spoil banks: Effects on a coastal marsh water level regime. Estuarine and Continental Shelf Science 24:599-609.

Tebeau, C. W. 1971. A History of Florida. Coral Gables, Florida: University Miami Press.

Turner, R. E. 1990. Landscape development and coastal wetland losses in the northern Gulf of Mexico. American Zoologist 30:89-105.

Turner, R. E. 1992. Coastal Wetlands and Penaeid Shrimp Habitat. Pp. 97–104, in R. H. Stroud, ed. Conservation of Coastal Fish Habitat.

Turner, R. E., and D. F. Boesch. 1987. Aquatic Animal Production and Wetland Relationships: Insights Gleaned Following Wetland Loss or Gain. Pp 25-39 in D. Hooks ed., Ecology and Management of Wetlands. Beckenham, Kent, UK: Croon Helms, LTD.

Turner, R. E., and D. R. Cahoon ed. 1987. Causes of Wetland Loss in the Coastal Central-Gulf of Mexico. Final report submitted to Minerals Management Service, New Orleans, Louisiana. Contract No. 14-12-001-30252. 0CS Study/MMS 87-0119.

Turner, R. E., N. N. Rabalais, and Z.-N. Zhang. 1990. Phytoplankton biomass, production and growth limitations on the Huanghe (Yellow River) continental shelf. Continental Shelf Research 10:545-571.

Turner, R. E., and Y. S. Rao. 1990. Relationships between wetland fragmentation and recent hydrologic changes in a deltaic coast. Estuaries 13:272-281.

Turner, R. E., and N. N. Rabalais. 1991. Water quality changes in the Mississippi River this century and implications for coastal food webs. Bioscience 41:140-147.

Turner, R. E., and N. N. Rabalais. 1992. Geomorphic variability of United States estuaries. Unpublished manuscript.

Turpin, D. H., and P. J. Harrison. 1980. Cell size manipulation in natural marine, planktonic, diatom communities. Can. J. Fish. Aquat. Sci. 37:1193-1195.

Valiela, I. 1984. Maline Ecological Processes. New York, New York: Springer Verlag.

Vollenweider, R. A., and J. Kerekes. 1980. The loading concept as a basis for controlling eutrophication: Philosophy and preliminary results of the OECD programme on eutrophication. Progress in Water Technology 12:5-38.

Weller, M. W., and L. H. Fredrickson. 1974. Avian ecology of a managed glacial marsh. Living Bird 12:269-291.

Wiegert, R. G., A. G. Chalmers, and P. F. Randerson. 1983. Productivity gradients in salt marshes: the response of *Spartina alterniflora* to experimentally manipulated soil water movement. Oikos 41:1-6.

8

Coastal Wetlands: Multiple Management Problems in Southern California

Joy B. Zedler
San Diego State University
San Diego, California

INTRODUCTION

California's 1973 Coastal Act was one of the nation's earliest attempts to plan for the coexistence of multiple coastal users. However, lack of support from recent governors, budget cuts, and intense population pressure have eroded California's status as a leader in coastal zone management. Multiple uses are now resulting in multiple conflicts (see Attachment 8.1), and estuarine wetlands are particularly threatened. The federal government has shown little interest in California estuaries. In fact, only two have been included in the Environmental Protection Agency's National Estuary Program: San Francisco Bay and Santa Monica Bay, and only four of the 19 National Estuarine Research Reserves are on the Pacific Coast. Of these, Elkhorn Slough and Tijuana Estuary are in California. The limited interest in Pacific coast wetlands extends to research support as well. It has been suggested that management models can be based on East Coast research and relationships and then modified to fit the west coast (Sutherland, 1991). This idea needs to be questioned.

Southern California estuaries have several unique qualities. The estuaries are small and isolated. The variability of various environmental factors (annual rainfall, timing of rainfall, storm intensity, and stream flow) is very high. Catastrophic events have lasting impacts on coastal wetlands. For example, Mugu Lagoon (near Santa Barbara) recently lost 40 percent of its low-tide volume due to flood-deposited sediments (Onuf and Quammen, 1983). Periodic El Niño events raise sea levels and increase storm frequencies. Coastal dunes are sometimes washed into the estuaries, especially where stabilizing vegetation has been denuded. At Tijuana Estuary, the shoreline has retreated 300 ft since 1852, with major erosion during the 1983 El Niño storms (Williams and Swanson, 1987).

Hydrologic features are also unusual. Freshwater discharge greatly influences the accumulation of sand from long-shore sediment transport processes in southern California. Estuarine inlets have a tendency to close, and the size of the tidal prism determines their ability to stay open to tidal flushing. Where watersheds are highly modified (disturbed soils and vegetation), erosion and sediment inflows can greatly reduce tidal prisms. Increased freshwater inflows cause native salt marsh vegetation to be replaced by brackish invaders (Zedler and Beare, 1986).

Finally, southern California has lost most of its coastal wetland habitat. In California as a whole, 91 percent of the wetland area (coastal plus inland) has been converted to other uses; this is the nation's highest loss rate (Dahl, 1990). On the coast, only about a fourth of the historic acreage is left, and much of it is in San Francisco Bay. Most of the 26 wetlands in southern California have some protection as habitat reserves. However, all have been reduced in size and are disturbed to various degrees. In the San Diego area, salt marshes have declined drastically. The acreage of tidal salt marsh in Tijuana Estuary, San Diego Bay and Mission Bay is only 13 percent of its historic area (San Diego Unified Port District, 1990). With all these habitat losses and damages, biodiversity is at risk. The state of California recognizes 10 coastal animal species as endangered or threatened with extinction (Department of Fish and Game, 1989). The California Native Plant Society considers 17 coastal wetland plants as rare.

This paper discusses two problems in southern California, both of which have aspects that are unique to the region. The first concerns wastewater management. Because municipal water supplies are imported from well outside the region, the release of treated effluent to streams threatens the hydrologic regime of coastal rivers and downstream estuaries. The second problem is mitigation. The region lacks the sites that are needed for mitigation projects, and there are no proven methods for replacing habitats used by endangered species. The paper ends with a consideration of the adequacy of the research base for dealing with these issues.

The Wastewater Issue

The more than 100 estuaries along California's 1100-mile-long coast receive streamflows in pulses, due to the region's Mediterranean-type climate with winter rainfall and summer drought. Under natural conditions, it is likely that streams had minimal flow in summer. In the San Diego region, dams reduce winter streamflows, and wastewater discharges increase summer streamflows, to coastal estuaries. Filling to build roads across the estuaries has reduced tidal prisms and increased chances of inlet closure. In general, the impact of development has been to decrease tidal influence and increase freshwater inflow, both by increasing the volume of fresh water discharged to the coastal wetlands and by prolonging the period of stream flow.

This region continues to grow very rapidly; more than 85,000 people moved to San Diego in 1987, and growth rates were just as high in 1988 and 1989. Development is moving inland, and it is becoming more expensive to discharge wastes to ocean outfalls. It has been proposed that the wastewater be treated and discharged to coastal streams for reuse in irrigation downstream during the dry season. The California Regional Water Quality Control Board projects discharges of 10 million to 30 million gallons per day of treated wastewater for 10 coastal rivers over the next 25 years (San Diego Regional Water Quality Control Board, 1988). It is uncertain how much of the flow would reach coastal wetlands, but certainly during the wet season, the wastewater discharge would exceed irrigation demands.

It is now recognized that changing the hydrology from intermittent to continuous flows will affect coastal water bodies and endangered species habitat. The coastal wetlands are usually saline to hypersaline ecosystems. A concern that is peculiar to semi-arid regions is salinity dilution, which occurs when intermittent streams that normally provide seasonal fresh water to coastal lagoons become year-round rivers due to wastewater discharge.

Some effects of dry-season flows to coastal wetlands have been documented. We know that prolonged periods of freshwater influence can force the replacement of salt marsh habitat (which is endangered species habitat) to brackish marsh (which is not) (Zedler and Beare, 1986; Beare and Zedler, 1987). Continuous freshwater flows also eliminate the marine invertebrates and shellfish that are native to many coastal lagoons. At Tijuana Estuary, the numbers of fish and macroinvertebrate species have been reduced substantially since 1986; numbers of individuals have dropped by an order of magnitude; and size distributions are markedly altered—for clams, only young-of-the-year can be found, indicating that larvae are available to settle in the estuary, but rarely survive to reproductive age (Nordby and Zedler, 1991).

Experimental tests of the effect of salinity dilution on fish and invertebrates have demonstrated that low salinity causes mortality, especially of molluscs (Nordby, Zedler, and Baczkowski, unpub. data). Detailed experimentation with California halibut shows that growth of juveniles is impaired by lowered salinity and that impacts are greatest on the smallest and youngest individuals (Baczkowski, 1992). Thus, modifications to the seasonality of streamflow (i.e., the semi-arid hydrology) of the region are seen as significant impacts, beyond the more general problems of nutrients and toxic materials that are carried in wastewater.

Decisionmakers are aware of the negative impacts of year-round inflows, and plans are underway to recover much of the treated effluent downstream for use in irrigation. There would still be spills and excess water during the wet season. Unfortunately, the impacts of excess freshwater discharge, of greater volumes of freshwater inflow, and of increased nutrient loadings to coastal water bodies are only generally predictable.

Raw Sewage from Tijuana

The city of Tijuana includes large urban areas that are not on sewer, and wastes are discharged as raw sewage to Tijuana River. About 13 million gallons per day were released to Tijuana River and Tijuana Estuary between about 1986 and 1991 (Seamans, 1988; Zedler et al., 1992). As in other regions, wastewater inflows carry unwanted materials into estuaries. What is unique in this case is the high concentration of pollutants, because of both lower per capita water use (concentrated wastewater) and fewer controls on contaminant loadings (industrial discharges).

The nitrogen and phosphorus that enter the Tijuana Estuary are largely of wastewater origin. Mexican sewage contains over 25 mg/l nitrogen and greater than 10 mg/l phosphorus. We have shown that Tijuana Estuary is nitrogen limited, and that macroalgal blooms are stimulated by wastewater inflows (Fong et al., 1987). Studies of heavy metals in Tijuana Estuary showed that surface water samples contain mean levels of 69 ppb cadmium, 55 ppb chromium, 281 ppb nickel, and 321 ppb lead (Gersberg et al., 1989). The lead level is relatively high. The sediments of the estuary, which may act as a sink for heavy metals, contained up to 1.7 ppm cadmium, 25 ppm chromium, 14 ppm nickel, and 59 ppb lead. Hot spots of contamination do exist in the estuary.

Short-term solutions and long-term plans have been developed. In fall 1991, the raw sewage was diverted to a holding lake in the United States, held briefly, and then pumped to San Diego's sewage treatment plant during off hours. However, the pumps were shut down during rain storms in winter 1992 and failed for one week in May 1992. The short-term solution is a band-aid approach.

A long battle has been waged over who would pay for wastewater treatment at the border. The city of San Diego did not want to pay for treatment of "international waste". But the city did want a treatment plant that would serve new developments on the U.S. side. The federal government, in turn, did not want to pay for local infrastructures. The compromise was to build two sewage plants, a 25 millions of gallons per day plant to handle Mexico's sewage and a much larger plant to be built by the city of San Diego to treat local wastewater. To handle the effluent from both plants, a 12-foot-diameter outfall is being constructed to carry up to 300 millions of gallons per day of wastewater to the ocean. This outfall would cross Tijuana Estuary and damage a 200-foot-wide swath of endangered species habitat during construction. Mitigation is proposed. An alternative tunnel is being planned; the outfall pipe could go under Tijuana Estuary at greater construction cost. It is not clear that the estuarine biota could sustain the damages of either construction project, even with mitigation efforts.

Management and Policy Needs

The region faces dwindling water supplies and burgeoning effluent. The need for long-term solutions is obvious. Year-round reuse of water would obviate the need for a destructive ocean outfall. Year-round reuse would also solve problems both at the source (San Francisco bay Delta, where freshwater inflows are needed to sustain the biota of the bay) and at the disposal site. The drinking of wastewater that is produced and treated in California is permitted only if it has passed through a groundwater aquifer. The concern is apparently the potential for transmission of viruses. Further research on the safety and acceptability of total recycling is needed.

Second, treated wastewater could be used to construct wetlands. Freshwater wetlands could subsidize habitat for estuarine birds; at the same time, they would improve water quality entering the estuaries. In San Diego County, freshwater bulrush *(Scirpus validus)* wetlands have a particularly high capability for nitrogen removal, with greater than 90 percent reduction of total nitrogen at 5–6-day hydraulic residence times (Gersberg et al., 1986). Constructed wetlands are also capable of removing both bacterial and viral indicators of pollution with a removal efficiency of nearly 99.9 percent for poliovirus (vaccine strain; Gersberg et al., 1987).

Since augmented inflows are detrimental to the region's estuaries, these constructed wetlands could be engineered to discharge treated wastewater in pulses that would minimize negative impacts (i.e., salinity dilution) on the downstream estuary. Recent experimentation with pulsed-discharge regimes (alternating impoundment and discharge) demonstrated that both metal and nitrogen removal rates could be increased by twice-daily impoundment and discharge (Sinicrope et al., in press; Busnardo et al., in press). Several additional benefits might also occur. More nitrogen would be removed through enhanced denitrification; more metals would be immobilized through precipitation in the sediment; there might be fewer problems with mosquitoes due to the more dynamic hydrology. It appears that the problem of augmented freshwater inflows could be lessened (although not eliminated) by using pulsed-discharge wetlands to reduce the impact of salinity dilution and improve the quality of effluent entering the coastal wetlands. The potential for using constructed wetlands to manage wastewater in southern California needs to be explored.

THE MITIGATION ISSUE

The principal value of southern California's coastal wetlands is habitat and its role in maintaining biodiversity. Several species are dependent on our estuaries (including plants and animals, invertebrates and vertebrates, and both resident and migrant species). Three endangered bird and one plant species depend on coastal wetlands that cover less than 25 percent of their historic area and that are far from pristine. Despite laws that protect wetlands and endangered species, regulatory agencies still permit habitat alterations if mitigation plans promise compensation. Lost habitat is usually *replaced* by restoring disturbed wetlands, with a net loss of wetland acreage and often a decline in habitat quality (Zedler, 1991).

The National Environmental Policy Act [40 CFR Part 1508.20(a-e)] defines mitigation as avoiding, minimizing, rectifying, reducing, eliminating, or compensating for impacts to natural resources. Wetland filling is regulated by the Clean Water Act, Section 404, which requires a permit for the filling of, or disposing dredge spoil into, wetlands. Filling is allowed for water-dependent uses (e.g., port facilities) and where there is "no practicable alternative", providing that impacts are mitigated. Wetland mitigation usually involves restoration or enhancement of disturbed wetlands. Rarely does it involve construction of new wetland habitats from nonwetlands. Whether restoration, enhancement, and construction measures can preserve coastal diversity remains a major question.

Mitigation Projects in Progress

The ports of Los Angeles and Long Beach propose to fill 2400–2500 acres of nearshore habitat by the year 2020 to expand port facilities. Several mitigation projects have been proposed, and at least one (at Anaheim Bay/Seal Beach National Wildlife Refuge) has been implemented. A second (374 acres of dredging at Batiquitos Lagoon) has reached the final EIR/EIS stage. Environmentalists believe the dredging is excessive and that it suits the ports' needs for mitigating fish habitat more than the lagoon's need for enhancement. The project is currently in litigation.

Some projects are not water dependent, but permits are still possible in southern California. A current proposal by the city of San Diego is to relocate and expand an existing sewer pump station within an intertidal salt marsh. The specific site was recently shown to support the largest population of an annual plant *(Lasthenia glabrata)* that is considered sensitive. Mitigation would be proposed. However, the native distribution, population dynamics, habitat requirements, and reproductive characteristics of this rare plant have not been studied.

Highways are usually not permitted in wetlands, but most of southern California's coastal wetlands are interrupted by three roadways (Interstate Highway 5, Pacific Coast Highway, and the Santa Fe Railroad).   There are continual plans to widen these roadways, with associated impacts on the remaining wetland resources. Along San Diego Bay, the salt marsh was recently damaged by three federal projects: the widening of a freeway, a new freeway interchange, and a new flood control channel. The US Fish and Wildlife Service determined that three endangered species were jeopardized by the projects, and compensatory mitigation was required (see next section).

The basic problems with mitigation in southern California are that 1) there isn't enough coastal acreage to satisfy the demand for mitigation projects; 2) too much wetland habitat has already been lost, and several species are threatened with extinction; 3) even the most disturbed wetlands provide some support for threatened species, so that changing a degraded habitat into a mitigation site causes further negative impacts; and 4) we don't understand how these degraded wetlands function. In addition to these problems, the research that has been done to assess the functional equivalency of restored and natural wetlands indicates that we do not yet know how to recreate endangered species habitats. To date, the process of mitigation has been an attempt to offset losses, but the policy breaks down at several levels, including planning, site selection, and project implementation.

An Attempt to Compensate for Lost Endangered Species Habitat

The San Diego Bay mitigation was a habitat conversion. Disturbed high marsh/transition was converted to low marsh for a federally endangered bird (the light-footed clapper rail). Prior to excavation, the mitigation site may have supported the Belding's Savannah sparrow, a bird on the state endangered list. No biological inventory was required or conducted to document existing values of the mitigation site.

Research is continuing to assess the functional equivalency of restored and natural wetlands of San Diego Bay. In 1985, eight salt marsh islands were constructed as habitat for an endangered bird. Over a two-year period (1987–89), eleven attributes of the mitigation site were compared with those at an adjacent natural marsh. There were deficiencies in soils (Langis et al., 1991), plant growth (Zedler, in press), and marsh invertebrates (Rutherford, 1989). Sampling of soils and vegetation continued through 1992, and improvements were minimal, giving little evidence that the site will eventually support the target species, the light-footed clapper rail.

Compared with reference marshes, the sediment was sandier and had little organic matter. With less soil organic matter, there was less energy and nitrogen for microbial mineralization and less energy for nitrogen fixation. With lower nitrogen inputs, plant growth was limited and foliar nitrogen was lower. With lower plant production and lower-quality plant biomass, the detrital food chain was probably impaired, as indicated by lower abundances of invertebrates in the epibenthos. Five years after construction, the best sites (areas with highest plant cover) provided less than 60 percent of the functional value of the natural reference wetland (Zedler and Langis, 1991). More recent research shows that canopy architecture (cordgrass height and density) differed for the planted marshes (which do not support clapper rails) and natural marshes. Tall plants are needed to support nesting and provide protection from aerial predators. Transplanted marshes have few plants over 60 cm, while most stems in natural marshes exceed 60 cm (Zedler, in press). This appears to be a major reason why the marsh islands are not yet used by the light-footed clapper rail.

It should be possible to accelerate development of these ecosystem processes using scientific knowledge and experimentation at existing mitigation projects. Current policies are not sufficient to protect against extinction. Research is underway to find soil amendments and enrichment schedules that will produce taller, denser cordgrass in a shorter period of time. Preliminary experiments with straw, alfalfa, and inorganic nitrogen fertilizers show that nitrogen addition can improve plant growth (Gibson, 1992). However, after 2 years, the canopy architecture is not yet equivalent to that of natural marshes. Continued research and repeated applications seem to be necessary.

Management Policies and Issues

It has not yet been shown that damages to endangered species habitat can be reversed or that lost wetland values can be replaced. Replacement of functional values is slow and incomplete. Yet, policies have not been changed to reflect the inadequacy of mitigation projects. Permits are still being granted with the promise that habitat can be replaced. Several federal policies are not appropriate for the southern California situation.

Mitigation priorities: Federal mitigation policy (Environmental Protection Agency and Corps of Engineers Memorandum of Agreement) recommends that restoration be given priority over the creation of new wetlands from upland. This policy makes sense in some places, such as prairie potholes that are drained and farmed and no longer function as wetlands. Restoration of former potholes is more likely to provide the correct hydrology than excavation of potholes from natural upland. However, where damaged wetlands still perform critical functions, as in southern California, this strategy is doubly damaging–first, the restoration site is altered without knowing what existing values were lost; second, there is a net loss in wetland area.

Mitigation ratios and "net loss of acreage and function: "The recommendation that mitigators restore 2–4 times the area they damage is a good idea. However, it is not sufficient where endangered species habitats are concerned. Since even the disturbed wetlands have valued functions, the use of a 2:1 mitigation ratio (i.e., restoration of 2 acres of marsh for every 1 acre lost) does not fulfill the policy of no net loss of wetland area. Instead, there is a net loss of 1 acre of wetland area. Only the creation of wetland from non-wetland areas can replace lost wetland acreage. Lower functional value of restored or created wetlands does not compensate for lost endangered species habitat. Even if 3:1 or 4:1 compensation is required, a larger area of unusable habitat will not replace the functional value of one acre that is critical to the endangered population. At the very least, agencies should require assessment of the functioning of the mitigation site prior to and after improvements, plus *up-front* mitigation, with *success* achieved and documented prior to destruction of the development site.

Sediment removal: Most restoration projects in southern California involve excavation of sediments that have accumulated from coastal watersheds or former fills. However, sediments may ultimately be needed to offset sea-level rise. Most agencies lack policies that require consideration of accelerated rates of sea-level rise in their long-term planning.

Dredge spoil disposal: Off-site disposal of fine sediments is extremely costly and may be environmentally damaging to the disposal site. A proposed solution for disposal of fine sediments at Batiquitos Lagoon is to bury them in situ, first excavating the underlying sand and using it for beach replenishment. However, this would extend the time period of the disruption of biota of Batiquitos Lagoon, and temporary stockpiles of spoils would affect nearby coastal areas. Regional plans for sediment disposal are needed, with an emphasis on finding beneficial uses of the material (e.g., capping toxic waste deposits).

Research policies: The National Oceanic and Atmospheric Administration's Coastal Ocean Program proposes to develop a conceptual model of estuarine habitat function based on East Coast models (Costanza et al., 1990). The objective is to "relate the location and extent of seagrass and salt marsh habitats to the production of living marine resources in an estuary or region" (Sutherland, 1991). Although this program has provided some support for the soil amendment experiment in San Diego Bay, most of the funding has been for research in East Coast and Gulf Coast habitats. The applicability of production models to southern California management issues concerning endangered species habitats is questionable. Research funding agencies need to recognize the unique attributes of Pacific Coast ecosystems and to reevaluate the geographic distribution of their funding efforts.

THE STATUS OF RESEARCH ON COASTAL
SOUTHERN CALIFORNIA ECOSYSTEMS

In November 1991, the California State Sea Grant Program sponsored a workshop on "Research Needs for Restoring Sustainable Coastal Ecosystems on the Pacific Coast" at the Estuarine Research Federation meetings in San Francisco (Williams and Zedler, 1992). The consensus was that ecosystem research on Pacific estuaries lags behind that on Atlantic and Gulf Coast estuaries by several decades. Even basic data on California estuarine wetlands (size, type, historic condition) are unavailable. Little is known of the habitat requirements of Pacific estuarine species, including plants, fish, and wildlife. For plant species that have been studied, such as *Salicornia virginica and Spartina foliosa,* we still do not have data on below ground dynamics. For these and other unstudied species, we lack data on dispersal mechanisms, reproductive strategies, and genetic structure. Estuarine food webs have not been elucidated, and feeding relationships have not been quantified. The research needs are numerous, as indicated by attendees at the recent national workshop (Table 2).

CONCLUSION

The uniqueness of Pacific coastal wetlands requires a regional approach to research and management. Whereas the nutrient content of fresh water entering East Coast and Gulf Coast estuaries needs to be controlled, in southern California the amount and timing of discharges must also be managed in order to maintain native vegetation and associated fauna. It is not sufficient for managers to worry only about the loss of fish and shellfish habitat, because endangered species are often jeopardized by wetland loss in southern California. Management models cannot be derived by extrapolation from data of East Coast and Gulf Coast estuaries, where inflows are more predictable and where plants and animals are more tolerant of brackish water.

REFERENCES

Baczkowski, S. 1992. The effects of decreased salinity on juvenile California halibut, *Paralichthys californicus*. M. S. Thesis, San Diego State University.

Beare, P. A., and J. B. Zedler. 1987. Cattail invasion and persistence in a coastal salt marsh: The role of salinity. Estuaries 10:165-170.

Busnardo, M. J., R. M. Gersberg, R. Langis, T. L. Sinicrope, and J. B. Zedler. In press. Nitrogen and phosphorus removal by wetland mesocosms subjected to different hydroperiods. Ecological Engineering.

Costanza, R., F. H. Sklar, and M. L. White. 1990. Modeling coastal landscape dynamics. Bioscience 40:911 07.

Dahl, T. E. 1990. Wetlands losses in the United States, 1780's to 1980's. Washington, D.C.: U.S. Department of the Interior, Fish and Wildlife Service.

Department of Fish and Game. 1989. 1988 Annual Report on the Status of California's State Listed Threatened and Endangered Plants and Animals. Sacramento, California: Resources Agency.

Fong, P., R. Rudnicki, and J. B. Zedler. 1987. Algal Community Response to Nitrogen and Phosphorus Loadings in Experimental Mesocosms: Management Recommendations for Southern California Coastal Lagoons. Technical Report to SANDAG, San Diego, California.

Gersberg, R. M., B. V. Elkins, S. R. Lynn, and C. R. Goldman. 1986. Role of aquatic plants in wastewater treatment by artificial wetlands. Water Resources 20:363-368.

Gersberg, R. M., S. R. Lynn, R. Brenner, and B. V. Elkins. 1987. Fate of viruses in artificial wetlands. Applied Environmental Microbiology 53:731-736.

Gersberg, R. M., F. Trindade, and C. S. Nordby. 1989. Heavy metals in sediments and fish of the Tijuana Estuary. Border Health V:5-15.

Gibson, K. 1992. The effects of soil amendments on the growth of an intertidal halophyte, *Spartina foliosa*. M.S. Thesis, San Diego State University.

Langis, R., M. Zalejko, and J. B. Zedler. 1991. Nitrogen assessments in a constructed and a natural salt marsh of San Diego Bay, California. Ecological Applications 1:40-51.

Nordby, C. S. and J. B. Zedler. 1991. Responses of fishes and benthos to hydrologic disturbances in Tijuana Estuary and Los Peñasquitos Lagoon, California. Estuaries 14:80-93.

Onuf, C. P. and M. L., Quammen. 1983. Fishes in a California coastal lagoon: Effects of major storms on distribution and abundance. Marine Ecology 12:1-14.

Rutherford, S. E. 1989. Detritus production and epibenthic communities of natural versus constructed salt marshes. M.S. Thesis, San Diego State University. San Diego Regional Water Quality Control Board. 1988. Staff report on stream enhancement and reclamation potential 1988 through 2015. San Diego. 35 pp.

San Diego Region Water Quality Control Board. 1988. Staff Report on Stream Enhancement and Reclamation Potential 1988 through 2015. San Diego, California.

San Diego Unified Port District. 1990. South San Diego Bay Enhancement Plan, Vol. 1, Resources Atlas. Oakland, California: California State Coastal Conservancy.

Seamans, P. 1988. Wastewater creates a border problem. Journal of the Water Pollution Control Federation 60:1798-1804.

Sinicrope, T. L., R. Langis, R. M. Gersberg, M. J. Busnardo, and J. B. Zedler. In press. Metal removal by wetland mesocosms subjected to different hydroperiods. Ecological Engineering.

Sutherland, J. 1991. NOAA Coastal Ocean Program Estuarine Habitat Program. Proceedings of a Workshop of Principal Investigators, Horn Point, Maryland. November 1991 Draft. Silver Spring, Maryland: National Oceanic and Atmospheric Administration.

Williams, P. B. and M. L. Swanson. 1987. Tijuana Estuary Enhancement: Hydrologic Analysis. Oakland California: California State Coastal Conservancy.

Williams, S. L. and J. B. Zedler. 1992. Research Needs for Restoring Sustainable Coastal Ecosystems on the Pacific Coast. LaJolla, California: California Sea Grant College.

Zedler, J. B. 1991. The challenge of protecting endangered species habitat along the southern California coast. Coastal Management 19:35-53.

Zedler, J. B. In press. Canopy architecture of natural and planted cordgrass marshes: Selecting habitat evaluation criteria. Ecological Applications.

Zedler, J. B. and P. A. Beare. 1986. Temporal variability of salt marsh vegetation: the role of low-salinity gaps and environmental stress. Pp. 295-306 in D. Wolfe, ed., Estuarine variability. New York, New York: Academic Press.

Zedler, J. B. and R. Langis. 1991. Comparisons of constructed and natural salt marshes of San Diego Bay. Restoration & Management Notes 9(1):21-25.

Zedler, J., C. Nordby, and B. Kus. 1992. The ecology of Tijuana Estuary: A national estuarine research reserve. Washington, D. C.: National Oceanic and Atmospheric Administration Office of Coastal Resource Management, Sanctuaries and Reserves Division.

Attachment 8.1 Major Uses of the Southern California Coast

Ports: The ports of Los Angeles and Long Beach occupy San Pedro Bay. To expand their role as the major Pacific Rim shipping center, they propose to fill 2400 more acres of shallow subtidal habitat by the year 2020.

> **Problems:** Nearshore fish habitat will be filled. Mitigation to compensate for lost habitat is mandated, but there are no nearby sites where compensatory habitat restoration or construction can occur, because all the historic sites have been filled and urbanized.

Marinas: Many former wetlands have marinas. There is constant pressure to increase the number of boat slips for San Diego's growing population and its substantial tourism industry. America's Cup generated further expansion.

> **Problem:** Marina development impacts eelgrass beds and associated fisheries (e.g., California halibut). No studies have documented the functional value of natural eelgrass beds or of mitigation sites where eelgrass has been transplanted. Since so much eelgrass habitat has been destroyed, and since eelgrass is a clonal species, transplanted eelgrass beds may lack genetic diversity.

Urbanization: Most of the coast above mean high water is urbanized. Were it not for a military base (Camp Pendleton), the 120-mile coast between Los Angeles and San Diego would be one continuous urban strip. The San Diego region now has 2.5 million people, with approximately 2 million more in adjacent Tijuana, Baja California. The growth rate is variable for the San Diego area, but newcomers averaged 85,000 per year from 1987-89.

> **Problems:** There is no buffer zone between urban areas and coastal habitats. Urban runoff degrades coastal water bodies; noise, lights, and human activities occur immediately adjacent to endangered species habitats. There is constant pressure to *use* wildlife preserves.

Military bases: San Diego grew up around the Naval Base on San Diego Bay. There is also a Marine Corps Recruit Training Depot on San Diego Bay and a Marine Base (Camp Pendleton) in northern San Diego County. There are military airfields at Tijuana Estuary (helicopters) and Miramar Naval Air field (jets), which is just north of San Diego.

> **Problem:** It is not known whether coastal military bases are releasing contaminants. Contaminants from anti-fouling paints (e.g., tributyltin, copper) are known to be a problem at the Navy harbor.

> **Problem:** San Diego Bay must be dredged to maintain ship channels.

Problem: Helicopters practice about 950 takeoffs and landings per day, with flights directly over the Tijuana River National Estuarine Research Reserve.

Airports: Tijuana, San Diego, Long Beach, Los Angeles, and Santa Barbara all have commercial airports. San Diego's airport is entirely surrounded by high-cost housing, commercial, and military land uses. Various alternatives are under consideration for major expansion.

Problems: Expansion in situ would encroach on the Marine Corps Recruit Depot and increase noise levels for nearby residents. Relocation to Miramar Naval Airfield would interfere with military activities. Locating a new airport adjacent to the international border (adjacent to Tijuana's airport) would increase flights over Tijuana and have planes taking off and landing over a National Estuarine Research Reserve. All U.S. users of the airport live north of the latter site; travel to and from the airport would be maximized, with most people having to drive through San Diego to reach the airport.

Agriculture: Minimal agricultural efforts are carried out along the coast; however, there are agricultural activities in the Tijuana River Valley, several areas of floriculture inland of coastal lagoons, and vegetable farming on the marine terrace at Camp Pendleton.

Problem: Non-point source pollutants and irrigation runoff flow into coastal wetlands. Algal blooms and fish kills are possible impacts.

Recreation: Sandy beaches offer many recreational opportunities. Mission Bay Park is billed as the "world's largest urban water-recreation park" (1,888 acres). Formerly shallow subtidal and wetland habitat, the park was constructed by dredging embayments and building islands. The area now supports waterskiing, jet skiing, sailing, rowing, canoeing, kayaking, swimming, sunbathing, and passive nature appreciation. At Tijuana Estuary, horse trails are heavily used by equestrian clubs and rental operations.

Problems: Incompatible uses are crowded into small areas. Noisy boats and jet skiers have negative impacts on those seeking quiet. Accidents occur in Mission Bay, jet skiers and boaters collide; at Tijuana Estuary, hikers get kicked by horses. It is not easy to eliminate incompatible users. A model airplane club that was located within the Tijuana River National Estuarine Research Reserve was ruled as incompatible when planes repeatedly crashed at an experimental research site. But it took over five years and considerable expense to evict them, even though their lease had expired.

Research and Education: These activities take place at several habitat remnants along the coast. Visitor centers occur at Tijuana Estuary, San Diego Bay, and at some of the north county lagoons. Tijuana Estuary is a research reserve, but it is managed by the California Department of Parks and Recreation, which does not have a research mandate.

Problem: Intensive use is damaging to native habitats and species; yet trails are desired for interpretive purposes. Most habitats are used by sensitive species, so there is no good place for interpretive or recreational facilities. Blinds are often suggested, but homeless people and undocumented migrants on their way north are attracted to such structures.

International Border: Every day, hundreds of undocumented immigrants cross the border at Tijuana, Mexico. The traffic flows on many undesignated paths, a number of which cross through or near endangered species habitat, such as the beach areas where California least terns nest.

Problems: Many of the immigrants walk, wade, and swim through Tijuana Estuary. They damage the habitat, as does the Border Patrol force, whose job it is to pursue and arrest them. Nests of endangered birds are damaged.

Gravel and sand extraction: Coastal rivers have been mined for both sand and gravel. These materials are highly valued in a region undergoing rapid development. An extraction company has developed an ambitious proposal to remove the surface 180 feet of the highlands [at the U.S.-Mexico border, sell the sand, and crush the cobbles to make gravel].

Problem: These highlands are an important buffer between Tijuana River Valley and the metropolis along the Mexico border. Eliminating these bluffs would have serious impacts on the integrity of the national estuarine research reserve. The long-term extraction period (> 10 years) would disrupt endangered species habitat throughout the valley. It is estimated that there would be more than one truck per minute transporting sand or gravel, plus the construction and operation of a rock crusher.

Coastal reserves: Coastal wetlands makeup a small but important acreage along the coast, since they support many species that are threatened with extinction. Several sites have been set aside for various conservation purposes. Tijuana Estuary is a national estuarine research reserve; it includes a state park and a national wildlife refuge for endangered species. At San Diego Bay, the Sweetwater Marsh is a national wildlife refuge. At Mission Bay, the Kendall-Frost Reserve is owned by the University of California and is set aside for research. North of San Diego, there are several coastal lagoons that are ecological reserves. The California Department of Parks and Recreation owns Los Peñasquitos Lagoon; the county of San Diego owns San Elijo Lagoon; and the California Department of Fish and Game owns Buena Vista Lagoon. Private holdings include large parts of Agua Hedionda Lagoon (San Diego Gas & Electric) and Ballona Wetland near the Los Angeles Airport (250 acres, Maguire Thomas Partners).

Problem: No single agency manages the region's wetlands. Yet migratory birds and mobile fishes and invertebrates are often the management target. Several species that are unique to the region are threatened with extinction. Tijuana Estuary supports 24 sensitive plant and animal species, yet the management of these populations is not easily coordinated within the region.

Problem: California's 91 percent loss of historic wetland area indicates that further extinctions will occur. E. O. Wilson's island biogeography model predicts that 50 percent of the species will be lost when 90 percent of the habitat is eliminated.

Problem: It is extremely costly to purchase wetland remnants for the public. The Famosa Slough was recently purchased by the city of San Diego (20 acres for $3.5 million) for a nature reserve and for public interpretation. This site had a development plan but it would probably not have received a Section 404 permit. Near Santa Barbara, public agencies paid over $340,000 per acre to acquire a 4-acre parcel adjacent to Carpinteria Marsh for purposes of restoration. This filled site was developable land.

Problem: Proposed changes in the wetland delineation guidelines would seriously impair habitat conservation efforts. The higher elevations of coastal marshes support rare and endangered plants (the salt marsh bird's beak) and insects (mudflat tiger beetles). With increased rates of sea-level rise, the upper marsh and transition habitats must be available for the landward migration. The Environmental Protection Agency calculates that a half meter rise by the year 2100 is probable—this would eliminate 65 percent (6441 square miles) of the wetlands of the contiguous United States.

Problem: Restoration plans are designed to meet the needs of mitigators, rather than what may be most functional at the site or most needed in the region. No program has been developed to assess the quality of each habitat type in the region, to assess losses by habitat type, or to determine the habitat needs of the region. Almost all mitigation planning is done on a piecemeal basis.

Problem: Much of the wetland area that remains in the region is publicly owned, but no single agency owns all the reserves. Management is accomplished on a site-by-site basis.

Attachment 8.2 Research Priorities for Sustainable Pacific Estuaries

This list is divided into four subject areas of equal importance: conservation of biodiversity, physical processes, water quality, and restoration. The categories were numerically ranked by workshop participants using 1 as the highest and 3 as the lowest priority. The tally below represents the mean ranking by fifteen respondents.

Conservation of Biodiversity Research Needs

Mean Rank Value	Categories

1.27	Habitat function determinants
	    structural (marsh edge, canopy height)
	    functional (productivity, trophic support)

1.30	Habitat requirements/habitat specificity of organisms
	    primary determinants of habitat utilization (trophic/reproduction requirements)
	    structure (e.g., habitat heterogeneity, canopy height)
	    function (e.g., productivity)

1.42	Population dynamics
	    genetic structure and diversity
	    minimum viable population sizes
	    community development processes (rates, rate-limiting processes)
	    below ground vegetation processes

1.89	Linkages between communities and habitats

1.90	Trophic dynamics
	    food web analysis
	    emergent insect communities

2.02	Exotic species biology
	    dispersal mechanisms
	    competitive effects
	    trophic effects

2.09	Habitat inventory
	    determination of estuarine acreage and habitat types
	    community profiles on sites with long-term data base

<u>2.10</u>   Endangered species biology

<u>2.65</u>   Effects of rare events

Physical Processes Research Needs

<u>1.23</u>   Hydrology
                effects of altered hydrology
                effects of vegetation
                effects of marsh morphology (channel vs. overmarsh flow)
                effects of alternating wet-dry cycles
                models

<u>1.78</u>   Erosion/accretion responses
                role of organic vs. inorganic matter in accretion
                integration with hydrological effects
                sediment supply processes

<u>1.97</u>   Model of salinity dynamics (modal & extreme)

<u>2.22</u>   Effects of anticipated sea level rise

<u>2.36</u>   Marsh morphology
                role of extreme events
                comparisons between marsh types

Water Quality Research Needs

<u>1.51</u>   Nutrient dynamics
                process rates
                budgets
                organic matter accumulation and decomposition rates
                effects of alternating wet-dry cycles
                effects of altered hydrology

<u>1.53</u>   Criteria for vegetation

<u>1.71</u>   Urban runoff

<u>1.81</u>   Impacts of development

<u>1.90</u>   Treatment strategies

Restoration Research Needs

<u>1.17</u>   Inventory of projects and monitoring

<u>1.29</u>   Habitat architecture
             habitat size to sustain minimum viable population sizes and functionality
             habitat heterogeneity
             landscape linkages and corridors
             buffer zone requirements

<u>1.59</u>   Site selection criteria
             identification of potential sites
             consideration of regional habitat biodiversity
             urban problems

<u>1.65</u>   Monitoring and evaluation of success
             assessment and standardization of functionality evaluation criteria
             assessment of appropriate temporal scales of monitoring
             assessment criteria for urban projects where no natural sites remain for
                 comparison
             assessment of structure (e.g., canopy height) as surrogates for function

<u>1.94</u>   Methodology
             identification of desired initial conditions
             establishment of desired initial conditions
             independent tests of design strategies
             acceleration of functional development trajectory
             incorporation of effects of rare/stochastic events in design

<u>2.48</u>   Economic evaluation of adequacy of mitigation/restoration options as
             compensation for loss

Source: Williams and Zedler, 1992.

9

Coastal Pollution and Waste Management

Jerry R. Schubel
The State University of New York
Stony Brook, New York

"The Future Ain't What It Used To Be."
Yogi Berra

INTRODUCTION

This paper was prepared as a background paper for the National Research Council's Commission on Geosciences, Environment, and Resources Retreat on "Multiple Uses of the Coastal Zone in a Changing World". In it I describe the major problems facing the coastal zone throughout the world and in the United States and review some of the priorities identified by the research and environmental management communities.

WHAT ARE THE MAJOR PROBLEMS OF THE WORLD'S COASTAL OCEAN?

The Joint Group of Experts on Scientific Aspects of Marine Pollution (GESAMP), an advisory group to the United Nations, periodically assesses the problems of the world ocean. In their most recent report (GESAMP, 1991) they pointed out that while human *fingerprints* are found throughout the world ocean, the open ocean is still relatively clean. However, there are serious problems in the coastal ocean. The report states:

> In contrast to the open ocean, the margins of the sea are affected almost everywhere by man, and encroachment on coastal areas continues worldwide. Irreplaceable habitats are being lost to the construction of harbors and industrial installations, to the development of tourist facilities and mariculture, and to the growth of settlements and cities.... If left unchecked, this will soon lead to global deterioration of the marine environment and of its living resources.

GESAMP (1991) summarized the major problems of the world ocean as

- nutrient contamination;
- microbial contamination of seafood;
- disposal of debris (particularly plastic debris);
- trace contaminants such as lead, cadmium, and mercury when discharged in high concentrations;
- occurrence of synthetic organic compounds in sediments and in predators at the top of the marine food chain; and
- oil in marine systems, mainly the global impact of tar bails on beaches and the effects of spills in local sheltered areas.

They added that radioactive contamination is a public concern. They did not consider the last two items above to be particularly important globally. In the summary of their findings, they stated:

> We conclude that, at the start of the 1990s, the major causes of immediate concern in the environment on a global basis are coastal development and the attendant destruction of habitats, eutrophication, microbial contamination of seafood and beaches, fouling of the seas by plastic litter, progressive build-up of chlorinated hydrocarbons, especially in the tropics and subtropics and accumulation of tar on beaches.

> ... not enough attention is being given to the consequences of coastal development, ... actions on land continue to be taken and executed without regard to consequences in coastal waters.

The GESAMP assessment is a global assessment of the entire world ocean and its coastal component. It's clear that the group's concern for the future of the world ocean is concentrated on the threats to the margins.

WHAT ARE THE MAJOR PROBLEMS OF THE U.S. COASTAL OCEAN?

Each year, the 23 coastal states, jurisdictions, and interstate commissions must report, for their estuarine waters, degradation that has reached the point that estuarine areas no longer fully support designated activities. In the most recent state Section 305(b) report to the U.S. Environmental Protection Agency, the 23 coastal states, jurisdictions, and interstate commissions reported that

- nutrients accounted for 50 percent[1] of the total impaired area of estuaries;
- pathogens accounted for 48 percent of the total impaired area; and
- organic enrichment/low dissolved oxygen accounted for 29 percent of the total impaired area.

The states cited municipal wastewater discharge as the most extensive single source of pollution to their estuarine waters. It accounted for 53 percent of the total impaired area. Non-point sources may have been underrepresented in the assessment.

It is clear that the problems of the U.S. coastal ocean, and the causes of those problems, are similar to those of the coastal zone of the rest of the world. The first order problems are eutrophication, pathogens, and habitat destruction. All are caused primarily by an increasing population and its waste disposal practices and by changing land-use patterns.

POPULATION AND ITS EFFECTS

The earth's population is now estimated to be nearly 5.5 billion and is projected to grow to more than 10 billion by the year 2050. Throughout the world, approximately half of all people live in coastal regions.

The increasing world population and the preferential settlement in coastal regions will only exacerbate the problems of the coastal ocean. Since 95 percent of the projected population growth will come in developing countries--countries with little or no infrastructure to manage human and industrial wastes--the most serious coastal-zone problems will be in developing countries.

Throughout the United States, nearly half of the population lives within 50 miles of the coasts of the oceans and the Great Lakes. Population in U.S. coastal areas has increased by about 30 million people over the last three decades, and this growth accounts for almost half the total U.S. population increase over that period. The U.S. coastal population is expected to continue to increase, although at reduced levels (Culliton et al., 1990). By the year 2010, the coastal population of the United States is projected to increase by almost 60 percent. Within coastal regions, people will continue to cluster near estuaries.

Estuarine and coastal areas not only are among the nation's most populous areas. They also are among the nation's most densely populated areas. Population densities are highest in the counties of the northeast and Pacific regions of the United States, which together account for 28 percent of the nation's total population. The northeast region, which extends from Virginia to Maine, is the most densely populated of the five regions (northeast, southeast, Great Lakes, Gulf of Mexico, Pacific). It contains 18 of the 25 most densely populated counties in the entire United States, and six of the nation's seven leading states in coastal county population. The distribution of population in the United States is shown graphically in Figure 9.1.

---

[1]The percentages total more than 1000 percent because more than one stressor contributes to impairment of an area.

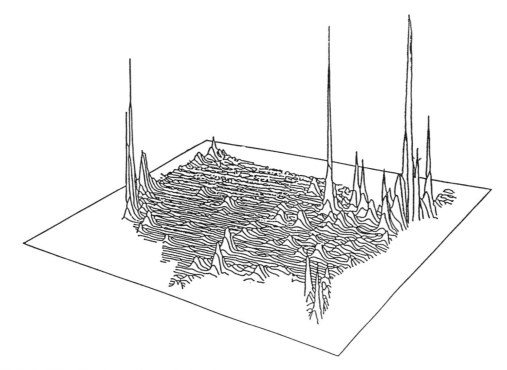

FIGURE 9.1 Distribution of population in the United States by region. (Source: Laboratory for Computer Graphics and Spatial Analysis, Harvard University)

As population in coastal regions grows, the coastal ocean loses. The greatest losses will occur in developing countries unless preventive measures are taken quickly.

GESAMP wrote in 1991:

> The exploitation of the coast is largely a reflection of population increase, accelerating urbanization, greater affluence and faster transport--trends that will continue throughout the world. Controlling coastal development and protecting habitat will require changes in planning both inland and on the coast, often involving painful social and political choices.

As the GESAMP report points out, protecting coastal habitat will require planning not only on the coast, but inland as well. For some estuaries, such as Chesapeake Bay, that planning must extend throughout much of the drainage basin. In others, such as Long Island Sound, the area of terrestrial influence is more constrained, and planning and management can be concentrated in the coastal zone. For each coastal system, the zone of influence of human activities needs to be identified and become the basic planning and management unit.

According to Goldberg (1990), tourism accounts for about 10 percent of the world's gross national product. In many developing countries, tourism is the main source of income. In coastal countries, much of the tourism is dominated by water-related activities. Some of these same developing countries are experiencing the world's most rapid population growth rates. Few have the resources--fiscal and technical--needed to construct, maintain, and operate the infrastructure needed to handle the wastes, particularly the human wastes, of their burgeoning populations. Typically, sewage is discharged raw into near coastal waters which causes a serious public health threat to bathers and to those who consume raw or partially cooked shellfish. The potential for major epidemiological outbreaks is high and growing.

There are other environmental impacts of discharging raw or improperly treated sewage into coastal waters, particularly into bays, estuaries, and lagoons. The added nutrients can produce eutrophic conditions leading to loss of submerged aquatic vegetation; to shifts in plankton assemblages; to degradation of coral reefs; and, in the extreme, to hypoxic or even to anoxic conditions. The most popular beaches and coastal environments and the tourists they attract are increasingly at risk.

The coastal areas at greatest risk are in developing countries. They can and should be identified now and steps should be taken to assist those countries in protecting them. Priority should be given to protecting those coastal areas that are still in good condition. Preventive environmental medicine is a far more effective and less costly strategy than restorative environmental medicine.

SOME TRENDS IN U.S. COASTAL WATERS

A widely held perception is that the coastal ocean is in rapid decline. Let's review quickly some of the data on contaminants and pathogens for U.S. estuaries.

Contaminants

A 1990 report by the National Oceanic and Atmospheric Administration's (NOAA) National Status and Trends (NS&T) Program summarizing six years of data on chemical contaminants in sediment and tissues states, "... it appears that, on a national scale, high and biologically significant concentrations of contaminants measured in the NS&T Program are limited primarily to urbanized estuaries. In addition, levels of those contaminants have, in general, begun to decrease in the coastal U.S.".

Even the higher levels in urbanized estuaries "... are generally lower than those expected to cause sediment toxicity, and among the NS&T sites, biological responses to contamination, such as liver tumors in fish or sediment toxicity, have not been commonly found... most contaminants measured in the NS&T Program may be decreasing. Except possibly for copper, there is little evidence that they could be increasing."

The chemical's measured in the NS&T Program are metals (Cd, Cr, Cu, Pb, Hg, Ag, and Zn) and organic compounds (tDDT, tCdane, tPCB, and tPAH). The NOAA Status and Trends sampling sites are intended to be representative; hot spots are avoided (NOAA, 1990).

Pathogens

The National Shellfish Sanitation Program (NSSP) classifies shellfish-growing waters to protect public health. It is a cooperative program involving states, industry, and the federal government. Since 1983, the NSSP has been administered through the Interstate Shellfish Sanitation Conference. The NSSP requires states to classify shellfish-growing waters according to approved protocols into four categories: Approved, Conditionally approved, restricted, and prohibited.

Data from 1985 and 1990 are summarized in Table 9.1. The pollution sources affecting shellfish-growing areas in 1990 are summarized in Table 9.2.

The data in Table 9.2 indicate the effects of coastal development on classification of shellfish-growing areas between 1985 and 1990. According to NOAA (1991) the largest increases in closures are attributed to urban runoff increasing from 23 to 38 percent of harvest-limited waters. The acreage adversely affected by septic systems increased from 22 to 37 percent. NOAA attributed the increasing effects of septic systems to the continuing growth of tourism and vacation homes. The impacts of boating rose from 11 to 18 percent.

Nutrients

I am unaware of any systematic summaries of the trends of nutrients in U.S. coastal waters. I expect that levels in many estuaries are increasing, primarily because of increased populations. In Long Island Sound, over the past 50 years the non-point- source input of nutrients from agriculture has declined, but the non-point-source input from creeping suburbanization has increased. Over the same period, the point-source inputs from New York City treatment plants has been relatively stable, but non-point- sources in coastal counties bordering the sound have increased significantly. Over-enrichment of Long Island Sound by nitrogen is considered by the Long Island Sound Study to be the most important hazard to the sound ecosystem. In 1991, New York and Connecticut signed a pact to cap nutrient inputs at 1991 levels and to work to decrease the input. To maintain nutrient inputs to the sound at 1991 levels--levels that are already too high--a significant investment will be required in the future--even in a region that now has one of the slowest population growth rates in the nation. Schubel and Pritchard (1991) estimated that in the year 2050, it would require an additional removal of 20-25 percent of the nitrogen to honor the 1991 cap.

The Top 10 Pollutants in Estuaries

Figure 9.2 shows the state Section 305(b) assessment of the top 10 offenders (pollutants) of the nation's estuaries in terms of their contributions to total impaired area. The sources of pollution are shown in Figure 9.3.

Table 9.1 Distribution of Classified Estuarine Waters, 1985 and 1990

Region	Approved 85	Approved 90	Prohibited 85	Prohibited 90	Conditional 85	Conditional 90	Restricted 85	Restricted 90
North Atlantic	87	69	10	29	1	1	2	1
Middle Atlantic	82	79	11	13	3	4	4	4
South Atlantic	75	71	22	21	3	4	<1	4
Gulf of Mexico	54	48	24	34	17	16	6	1
Pacific	42	53	40	31	18	11	1	5
Total	69	63	19	25	9	9	4	3

Percent Classified

(Source: NOAA, 1991)

Table 9.2 Pollution Sources Affecting Harvest-Limited Acregage, 1990[a,b]

	North Atlantic		Middle Atlantic		South Atlantic		Gulf of Mexico		Pacific		Nationwide	
	Acres	%	Acres	%	Acres	%	Acres	%	Acres	%	Acres	%
Point Sources												
Sewage Treat Plants	238	67	641	57	374	44	973	27	75	25	2,307	37
Combined Sewers	21	6	224	20	0	0	211	6	0	0	457	7
Direct Discharge	1	<1	84	7	5	1	920	25	6	2	1,015	16
Industry	21	7	223	20	180	21	522	14	129	42	1,077	17
Nonpoint Sources												
Septic Systems	91	26	123	11	288	34	1,763	48	57	19	2,322	37
Urban Runoff	75	23	655	58	290	34	1,276	35	110	36	2,412	38
Agricultural Runoff	5	3	130	12	233	28	301	8	41	13	718	11
Wildlife	19	7	112	10	306	36	1,115	30	39	13	1,597	25
Boats	55	17	353	31	146	17	507	14	47	15	1,113	18
Upstream Sources												
Sewage Treat Plants	2	1	104	9	9	1	1,174	32	45	16	1,334	21
Combined Sewers	0	0	5	<1	0	0	134	4	0	0	0	2
Urban Runoff	3	1	72	6	8	1	793	22	43	14	918	15
Agricultural Runoff	0	0	1	<1	0	0	435	12	0	0	436	7
Wildlife	0	0	28	2	35	4	210	6	0	0	273	4

a. Acres are times 1,000; % is percent of all harvest-limited acreage in region.
b. Since the same percentage of a shellfish area can be affected by more than one source, the percentages shown above cannot be added. They will not sum 100.

(Source: NOAA, 1991)

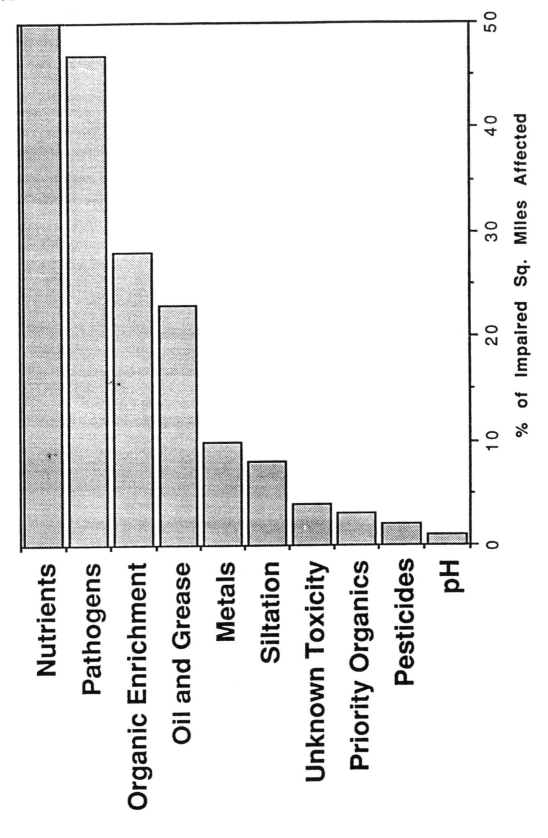

FIGURE 9.2 Top ten pollutants in estuaries.
(Source: USEPA, 1990)

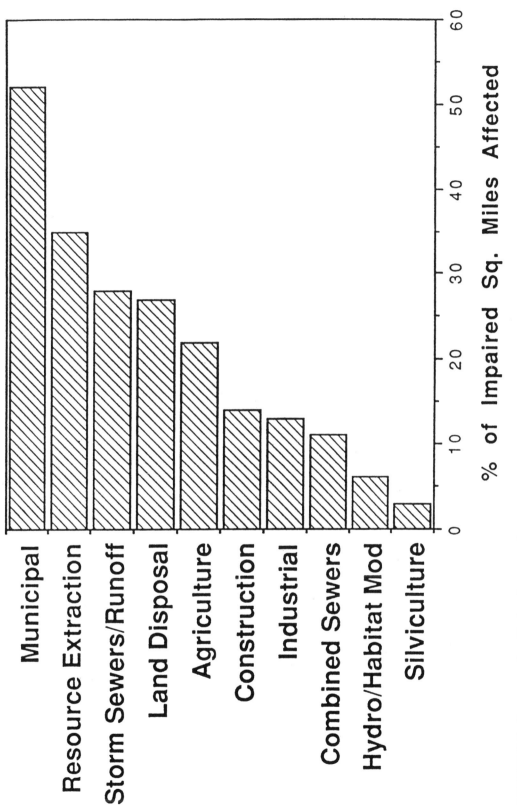

FIGURE 9.3 Sources of pollution in estuaries.
(Source: USEPA, 1990)

**One Person's List of the 11 Worst (Most Degraded) Estuaries
and Near Coastal Regions**

If pressed to come up with the *Big 11* of the nation's most degraded estuaries and coastal regions based on: (1) levels of pollutants in bivalves (clams, oysters, mussels) and sediments, (2) hypoxia/anoxia, (3) depleted and closed fisheries, (4) prevalence of fish diseases, (5) areas closed to shellfishing, (6) areas closed to swimming, and (7) warnings concerning consumption of fishery products, the following would make my *unranked* list of coastal areas of greatest concern:

- Boston Harbor
- Narragansett Bay
- Buzzards Bay
- Western Long Island Sound
- Baltimore Harbor
- Upper Chesapeake Bay
- Hampton Roads/Elizabeth River (Chesapeake Bay)
- Lower Mississippi and inner delta
- Galveston Bay
- San Francisco Bay
- Portions of Puget Sound

WHAT RESEARCH PRIORITIES HAVE BEEN IDENTIFIED?

The majority of the most serious problems of the coastal ocean are fairly well documented. There are few surprises. In this section, I consider briefly the extent to which research priorities reflect these problems.

Over the past two decades, there has been a series of workshops to identify the research *needs* for estuaries and near-coastal waters. Often these workshop retreats were held in idyllic spots; they always included many of the leading scientists. Whether the workshop was held on Block Island, Catalina Island, or Long Island, whether it was in North Carolina or in East Anglia (UK), the lists of research priorities were remarkably similar. This is not surprising: the problems of the coastal zone are pervasive and persistent, and many of the participants were repeaters. What is surprising is the lack of improvement in the richness with which the specific questions have been formulated and in the evolution of the research programs to attack them.

The results of some of these workshops are summarized in Table 9.3. The consensus on priorities is clear. If another workshop were held in 1992--and I'm not advocating it--the list would differ little. If a workshop were to be held, it could more profitably concentrate on a single priority issue of long standing, such as eutrophication, state the research problems more richly, and give more specific guidance for formulation of a research program to advance the level of our understanding. It should

Table 9.3 Estuarine Research Priorities, 1974-1990

Research Priorities	NSF 1974	NRC 1977	NERC 1982	CRC 1983	NRC 1983	Hudson River 1984	Sea Grant 1984	NASULGC 1986	NOAA 1986	ASLO 1990	MSRC 1990a	MSRC 1990b
• Sediment Management	X	X	X	X			X	X	X	X	X	X
• Behavior of Suspended & Dissolved Matter		X	X	X	X			X	X	X		X
• Resuspension; Remobilization; Role of Physics, of Biota	X	X	X	X						X	X	X
• Water Management							X	X	X			
• Circulation & Mixing	X	X	X		X			X		X		X
• Toxics & Other Contaminants		X		X		X	X	X	X		X	
• Eutrophication	X	X	X	X			X	X	X	X		
• Habitat, Fisheries Prod.				X	X	X	X	X	X	X	X	
• Behavior & Fate of Particulates		X	X					X	X	X		X
• Population, Community			X	X				X		X	X	
• Anoxia/Hypoxia		X		X				X	X			
• Monitoring			X	X		X		X				
• Coupling of 1° and 2° Production						X	X		X			
• Plumes & Fronts		X	X									
• Data Management				X				X				
• Sources of Carbon			X						X	X	X	
• Waste Management		X		X								
• Air-Water Exchange	X											
• Global Warming									X			

be structured to build partnerships with decisionmakers from the outset in order to utilize the new knowledge.

SOME POTENTIAL EFFECTS OF GLOBAL WARMING

While the major direct effects of global warming on the coastal ocean will not be on *coastal pollution and waste management*, there may be some indirect effects. And, since I have the floor...

Sea Level

Half of the world's population lives in coastal regions, many of which are already under stress. Eustatic sea level has been rising for the past approximately 18,000 years. Regionally, the rate of rise of sea level may be either greater or less than the worldwide average because of regional isostatic adjustments. An increase in the rate of rise of sea level because of global warming will have its greatest impacts on low lying coastal areas already subject to flooding. As much as 20 percent of the earth's population lives on lands that would likely be inundated or dramatically changed. Bangladesh and Egypt are among the world's most vulnerable nations to a rise in sea level. But, they are not alone.

An interesting example of a nation that would be impacted is the Republic of Maldives. This nation consists of an archipelago of about 1190 small islands, which lie approximately 6100 km southwest of Sri Lanka. Most of the nation rises only 2 m above sea level. In 1987 Maldive's President Maamoon Abdul Gayoom went before the United Nations General Assembly and described his country as *an endangered nation*. He pointed out that the Maldivians "did not contribute to the impending catastrophe... and alone we cannot save ourselves".

Tidal wetlands may be one casualty of an increase in the rate of rise of sea level. Wetlands--particularly youthful wetlands--are able to maintain themselves in a rising sea either by building vertically by trapping sediment and organic detritus or by moving landward. In many developed coastal areas, the lateral migration of wetlands has been halted by shoreline structures and by coastal construction. In others, the supply of sediments has been reduced because of better soil conservation practices and construction of reservoirs.

Titus (1990, 1991) stated that if current management practices continue and if sea level rises as projected, most of Louisiana's wetlands could be lost in the next century. These and other reports have indicated that a 1 m rise in sea level by the year 2100 could drown 25-80 percent of all U.S. coastal wetlands.

Saltwater intrusion into coastal aquifers and greater penetration of salt water into estuaries may threaten drinking water supplies.

Increased Frequency and Intensity of Storms

Because of the large concentrations of people in coastal areas, risks to life and property because of coastal storms are already high and will increase with population and sea-level rise. According to NOAA, a conservative estimate of the average economic costs of coastal hazards in the United States is about $2 billion/yr. However, Hurricane Hugo, alone, caused more than $9 billion in property damage and economic losses within the U.S. and its possessions.

If global warming causes a rise in sea level and increases the frequency and intensity of storm activity, flooding and storm damage to low-lying coastal areas will, of course, increase with an increased loss of property and human lives. Damage would be particularly great in delta regions of South Asia.

There will be other impacts of an increase in storm activity on coastal regions, most bad, a few perhaps good. Coastal infrastructure, such as sewage treatment plants, airports, power plants, and even the subways of some coastal cities, will be at greater risk. Increased storm activity also could increase the disturbance of contaminated sediments and the mobilization of contaminants. On the positive side, increased wind mixing of coastal waters by greater storm activity might alleviate the effects of hypoxia in some areas such as Long Island Sound and the New York Bight.

SOME OVERLOOKED PROBLEMS/PROCESSES

A number of coastal problems/processes have not received an appropriate level of attention. These include non-point sources, including the atmosphere; pathogens; eutrophication; and the manipulation of river discharges on coastal ecosystems. Because of limitations of space, and because land-derived non-point sources are beginning to receive far more attention, I will restrict my comments to atmospheric inputs and to the manipulation of river discharges.

You may be wondering how the latter relates to my assigned topic, *coastal pollution and waste management*. According to GESAMP (1991), "marine pollution means the introduction by man, directly or indirectly, of substances or energy into the marine environment (including estuaries) resulting in such deleterious effects as harm to living resources, hindrance to marine activities including fishing, impairment of quality for use of seawater and reduction of amenities." In this case, salt is the pollutant; it comes from the ocean, but man allows more of it to enter because of manipulation of the hydrologic cycle. As freshwater inflows are decreased, salt penetrates farther into estuaries, destroying low salinity habitat.

Manipulation of River Discharge: *A Looming*

One set of problems that has received too little attention and that may become more serious in the future is the effect of manipulation of river discharge on estuaries and near-coastal waters. If global climate change results in regional scale changes in precipitation patterns, in those areas where precipitation decreases, the coastal environment may be the big loser. When the value of water is high and when there is not enough to go around, as in California, the coastal environment has not competed successfully in the water allocation game.

The United States continues to have a voracious appetite for water. While it does not lead the world in any of the reported categories of water use (public, industry, electric cooling, and agriculture), in the aggregate the United States has the highest per capita water use and the highest total water use of all countries. China is second in total water use, and Canada is second in per capita water use.

A small number of rivers dominate the discharge of water to the world ocean. One river, the Amazon, accounts for more than one-third (34.6 percent) of the total water discharge of all the world's rivers. The Congo River ranks second with 6.9 percent of the total. Twenty-one of the world's rivers account for more than 90 percent of the total discharge; four of them account for more than 50 percent.

The human activity that has the greatest effect in reducing the discharges of water and sediment by rivers has been the construction of dams and reservoirs. Dams and reservoirs have also affected the pattern and timing of discharges. In Africa and North America, 20 percent of the total discharge is regulated by reservoirs. In Europe 15 percent is regulated, and in Asia--excluding China--14 percent is regulated. Only in South America and in Australasia are human impacts on river regimes relatively minor. According to Croome et al. (1976), "Some ten percent of the world's total stream flow now is regulated by men, and by the year 2000 it is probable that about two-thirds of the total discharge will be controlled."

While the prediction of Croome et al. may be an over-estimate--and I believe it is--the regulated fraction of the world's river discharge will increase and changes in regional precipitation patterns could have an influence.

The most intensive period of dam-building activity was between 1945 and 1971 when more than 8000 major dams were built outside of China (Beaumont, 1978). The year of peak activity was 1968 when 548 dams were commissioned. Beaumont's (1978) data do not include China which in 1982 accounted for more than 50 percent of all the world's dams, most of which were constructed after 1950 (Schubel et al., 1991). The United States ranks second in total number of dams, Japan third.

Reservoirs also trap sediment that would normally be carried downstream to coastal areas. Prior to construction of the Hoover Dam (1935), for example, the Colorado River discharged between 125-250 million $t.y^{-1}$ of sediment to the Gulf of California. In the decades after closure of the dam, the discharge dropped to only about 100,000 $t.y^{-1}$ (0.05 -0.1 percent of pre-dam levels; Meade et al., 1990).

Construction of dams on the Missouri River nearly eliminated the discharge of sediment from the Missouri to the Mississippi River--the Mississippi River's major source of sediment. Partly as a result of this, the sediment discharge of the Mississippi has fallen to less than half of what it was before 1953 (Meade et al., 1990).

The Aswan High Dam on the Nile River is perhaps the most striking example of the effects of a dam on the sediment and water discharges of a major river. After closing of the dam in 1964, the sediment discharges of the Nile to its delta dropped from an average of more than 100 million t.y^{-1} nearly to zero. The delta has been eroded and fisheries have collapsed.

The reductions in discharge of fresh water and sediment to estuaries and the reductions in the variability of freshwater inputs have effects on physical, chemical, and geological processes of estuaries and on their ecosystems. As competition for freshwater increases, the needs of estuaries will be weighed against the needs--real and perceived--of humans for water for drinking and domestic use, for agriculture, for cooling water, for electric generating stations, and for industry. In the absence of compelling arguments, estuaries will lose. They will be unable to compete successfully in the marketplace for freshwater unless the rules are changed to place a greater emphasis on the public trust doctrine and on the importance of preserving estuarine habitats.

Perhaps the Precautionary Principle is the place to begin. The Precautionary Principle can be stated in terms of the need to take a cautious approach to any actions that might degrade the environment and its living resources even before a causal link has been unequivocally established. The Precautionary Principle has to apply in all situations, not just in those where high priority activities are not threatened. If the Precautionary Principle were a guiding principle in the allocation of fresh water from the Sacramento-San Joaquin system in California, it is difficult to see how further diversions would be considered even in the absence of an unequivocal causal link between diversion and adverse effects on ecosystem values and functions in the low salinity portion of the estuary.

In the 1981 National Symposium on Freshwater Inflow, Rosengurt and Haydock (1981) stated "Direct experience and the published results of the effects of water development abroad, all point to the inescapable conclusion that no more than 25-30% of the natural outflow can be diverted without disastrous ecological consequences." Their observation was based upon studies of rivers entering the Azov, Caspian, Black, and Mediterranean Seas. In the same report, Clark and Benson (1981) state "Comparable studies on six estuaries by the Texas Water Resources Department showed that a 32% depletion of natural freshwater inflow to estuaries was the average maximum percentage that could be permitted if subsistence levels of nutrient transport, habitat maintenance, and salinity control were to be maintained." Again in that same report, Bayha (1981) indicated that results of studies by the Cooperative Instream Flow Service Group of the U.S. Fish and Wildlife Service "square well" with the observations of Rosengurt and Haydock.

The 25-30 percent criterion for maximum allowable reduction in natural riverflow does not have widespread acceptance among scientists or decisionmakers. According to Herrgesell et al. (1981) discharge of fresh water into San Francisco Bay has been reduced by approximately 50 percent since the 1800s. Other sources put the reduction at 70 percent. Some have predicted that inflows could be reduced to 10-15 percent of pre-diversion levels by the year 2000. Even with the major reductions that have already occurred, estuary managers and scientists face a formidable challenge in convincing the State Water Control Board that further reductions cannot be tolerated.

Clark and Benson (1981) suggested establishing optimal salinity regimes and associated hydrologic regimes within estuaries. Bayha (1981) pointed out that although estuarine needs are included among instream uses, few instream flow studies have actually incorporated an analysis of estuarine inflow requirements to ensure estuarine ecosystem values and functions.

The San Francisco Estuary Program is developing the scientific basis for a salinity standard to conserve low salinity habitat and living resources. The standard would take the form of an upstream seasonal limit for the position of the 2 percent near-bottom isohaline (Schubel et al., 1991). Even the discussion of a salinity standard has created concern.

The Atmosphere -- An Underestimated Source of Contaminants to the Coastal Zone?

The atmosphere may be underestimated as a source of a number of contaminants to coastal waters, particularly in urban areas such as Long Island Sound. While data specific to Long Island Sound atmospheric loadings are limited, preliminary estimates indicate that for a number of contaminants (Cu, Pb, Zn, PCBs, PAHs) direct atmospheric deposition on the sound may be of the same order of magnitude as the inputs from point and non-point sources. For example, analysis of atmospheric deposition rates of a variety of contaminants on high marshes bordering the Sound suggest that the atmosphere supplies (1) 90 percent of all Pb, (2) 35 percent of all Zn, and (3) 70 percent of all Cu supplied to the sound from all sources (Merkle and Brownawell, in press).

The implication is that for some urban coastal areas, the Clean Air Act may be more important than the Clean Water Act in reducing the levels of a number of contaminants.

ON THE NEED FOR NEW PARADIGMS

It should be clear that many problems of coastal areas are pervasive and persistent. Many have eluded solution. In developing countries, coastal pollution problems loom large. In both developed and developing countries, what are needed are new approaches to old problems. More of the same levels and kinds of research will produce only incremental improvements in our level of understanding of the causes of the problems and their effects, and in our ability to manage them. New approaches are needed that will take coastal marine science and management to their next Levels.

Science

One essential component of any successful approach--but only one--must be a stable, sustained program of unfettered research on coastal processes that is sufficiently attractive that it will capture the attention of the best minds in a variety of fields. It must be a combination of *big science*--multidisciplinary, multi-investigator studies that will tackle the next generation of coastal experiments and theory--and *small science*--science that will appeal to individual scientists.

We've always had the latter, although never enough to satisfy us; we've rarely-- if ever--had the former in coastal science, and never at the levels needed. The programs of basic research must encourage high risk research. Herbert A. Simon (1986) observed:

Science is an occupation for gamblers. Of course, journeyman science can be done without must risk taking, but highly creative science almost always requires a calculated gamble. By its very nature, scientific discovery derives from exploring previously unexplored lands. If it were already known which path to take, there would be no major discovery--and the path would most likely have previously been explored by others.

There has been too much parochial, journeyman science in the coastal ocean. Scientists in the coastal community need to emulate their deep water colleagues. They need to propose large, multidisciplinary projects that will attract teams of the best scientists from institutions across the country and around the world. There are a few encouraging signs; to mention two--the Land Margin Ecosystem Research Program and the proposed Coastal Ocean Processes (CoOP) Program.

The new CoOP Program is an exciting, multidisciplinary research program designed "to obtain a new level of quantitative understanding of the processes that dominate the transports, transformations and fates of biologically, chemically and geologically important matter on the continental margins." The CoOP Program prospectus (Brink et al., 1992) states that the *"The scientific results of CoOP will be useful in dealing with societal problems as well as purely scientific questions."* But will they? Could we shorten the time lag between advances in knowledge and applications to solve societal problems? More about that later.

New breakthroughs in coastal marine science will come, but one can't predict where, when, or by whom. We can, however, improve the conditions that nurture creativity and innovation.

Last year I was invited by the Estuarine Research Federation to present an historical overview of the evolution of estuarine physics. Because I believe many of the lessons apply to other areas of estuarine and coastal science, I want to comment on one section of that paper. It is the section that dealt with some of the factors that influenced the rate of evolution of estuarine physics.

During the 1950s and 1960s, there was a rapid evolution of estuarine physics; the development had been far more modest until then. In part this was due to the fact that it was in the 1950s that the physics of estuaries was first attacked in any serious and sustained way. Estuarine physics was virgin territory, and because of that, the probability of early explorations leading to major breakthroughs was high. Many of the problems were *zero-order* problems, problems that dealt with linear processes or with processes that were assumed to be linear.

The next generation of problems deal with non-linear processes and are far more complex and in some ways less attractive to scientists because of this. But, I am convinced that there was another reason for the rapid evolution of our knowledge of estuarine physics in the 1950s and 1960s. That reason was the strong, stable institutional support provided primarily by the Office of Naval Research and the Atomic Energy Commission.

That support enabled a few strong intellectual leaders to build and sustain research teams that included scientists--both experimentalists and theoreticians--engineers, technicians, and graduate students. Hypotheses were formulated, equipment was designed and built to make the critical observations, and field experiments were designed and carried out that utilized that instrumentation to test the hypotheses. The results were analyzed and interpreted, new insights resulted, new hypotheses were formulated, instruments were modified, and the next generation of field experiments was designed and carried out. The opportunities for all members of the team to *muck about* were rich. The opportunities to take chances and to fail were far greater than in today's funding climate. Progress was measured against a different *bottom line*.

In his excellent, recent book on creativity and problem solving, Kim (1990) makes this interesting observation:

> Studies of creativity in both science and art support the hypotheses that the likelihood of obtaining successful results does not vary significantly from one individual to another, nor among projects by a single individual. Rather the number of successes depends on the number of attempts that are made...

If Kim is correct, and I believe he is, he offers an additional reason why sustained institutional support resulted in major advances in our understanding of the physics of estuaries. This hypothesis of the importance of institutional support is consistent with Kim's Principle of Accelerated Failure: when the cost of failure is low, one should fail quickly and often. Kim asserts the obvious "to accelerate movement toward a final goal, it is necessary to take risks." Mechanisms that reduce the costs of failure encourage risk taking.

The last significant institutional support for estuarine research ended in the early 1970s. The shift away from institutional research support for *programs* to support for investigator-driven *projects* increased apparent efficiency but may well have contributed to a loss of effectiveness. Program managers and science administrators often confuse the two terms.

We may be seeing another spurt in the evolution of our understanding of the physics of estuaries. This one is driven by new instrumentation. It is clear that in research programs targeted at processes, scientists should conduct their research in the water body in which the processes they are interested in are revealed most clearly and most richly. This means that most such programs must be federally funded. New York will fund us to work in the Connecticut portion of Long Island Sound--sometimes--but they won't fund us to work in Massachusetts.

A program of fundamental research with these qualities is necessary if we are to make significant advances in our understanding of coastal processes and in our ability to manage coastal systems, but it is not sufficient. It must be combined with studies of coastal systems. While the processes may be the same in different coastal environments, the relative importance of those processes and their manifestations vary dramatically not only from one coastal system to another but often spatially and temporally within a single coastal system. Knowledge of the regional context is required for effective environmental management. You don't manage at a generic level--not in baseball or in coastal management.

A program of science in support of management, then, requires a combination of fundamental studies of processes with studies of coastal systems. While a program of sustained fundamental research in the coastal ocean and studies of specific coastal systems are necessary, even they are not enough if we are to conserve these valuable resources. We must develop new paradigms not only for coastal research but for coastal research in support of coastal management.

Coastal marine science must not only be good; it must be good for something. The coastal ocean is the most impacted part of the world ocean and the most politicized. The public's expectations for the coastal ocean are high, and their perceptions often do not closely track reality. If advances in knowledge of the coastal zone are not applied with little delay to the resolution of practical problems, the programs of fundamental research will be perceived to be failures, and it will be difficult to sustain support for them. In today's climate, some projects could be aborted by their very titles. It is unlikely today that many directors of coastal marine institutions would sign off as Don Pritchard did in the mid-1960s on a project of mine for a study of the "Effects of Dissolved Gases in An Old Woman's Gut" (Old Woman's Gut is a waterway in upper Chesapeake Bay).

Most coastal management problems are attacked by resource and regulatory agencies at the local or regional levels. As has already been pointed out, the advances in knowledge and understanding of scientific processes must be embedded in studies of individual coastal systems. A rich national tapestry of the coastal ocean that does not have detailed renditions of specific coastal systems will not be a useful map for guiding management of the coastal ocean.

At this point you may be saying to yourself, "Hasn't he every heard of the Environmental Protection Agency's National Estuary Program?" Yes, I have. I not only have heard of it, I've been involved in a few of its programs. The program has many merits, but it also has serious, fundamental flaws. The political forces are strong. Once an estuary is selected for inclusion in the program, the public's expectations run high. The expectations typically take the form of "Finally we can have one last round of studies, tie things together neatly in the development of a CCMP--a Comprehensive Conservation Management Plan--and take the prescribed actions that will protect our estuary for all time." We must change the public's mindset on the importance of sustained programs of fundamental research if we expect the public to share our conviction that we will have study estuaries and other coastal environments so long as they are important to society. The best way to accomplish this is to have more scientists involved in demonstrating the importance of their research in problem solving.

The pressures in National Estuary Programs to be inclusive are enormous. There are Science and Technology Advisory Committees, Citizens Advisory Committees, Management Committees, Policy Committees, Monitoring Committees, and Work Groups, and in some programs it has been necessary to create a Committee on Committees to try to keep track of the other committees. Roles are confounded and confused. Technical judgments sometimes are reached by consensus, not consensus among the scientific and technical community but consensus among the broader community. It might be well to remind them of Lewis Thomas' observation in his book *Late Night Thoughts on Listening to Mahler's Ninth Symphony*:

> There are some things about which it is not true to say that every man has a right to his own opinion. I do not have the right to an opinion on causality in the small world, or about black holes or other universes beyond black holes in the larger world, for I cannot do the mathematics. Physics, deep and beautiful physics, can be spoken only in pure, unaccented mathematics, and no other language exists for expressing its meaning, not yet anyway. Lacking the language, I concede that it is none of my business, and I am giving up on it.

I recently had the opportunity to review the draft document *U.S. Coastal Ocean Science, A Strategy for the Future*, which is being prepared under the Federal Coordinating Committee on Science, Engineering and Technology's Committee on Earth and Environmental Sciences. The document defines the initial steps in "...a strategic framework within which the Federal science agencies will work to improve the scientific basis for environmental decision making for the coastal ocean." It is a carefully crafted, well written, and comprehensive document that includes all the items in Table 3 plus a number of others.

Like all the other documents that have been prepared to develop the knowledge necessary to understand the coastal ocean so that we can protect it, the report is too much and too little. It is a catalog of problems and issues without enough information to show which of the problems and issues are most important. And the scientific questions in support of program goals represent, at best, only a modest improvement over previous statements. But there is a more serious and fundamental problem: the lack of coupling of science and scientists with management and managers. It is in this arena that NOAA has a unique role to play--not a bit part but a leading role. NOAA is the only federal agency that has a mandate for the coastal ocean that includes responsibilities for basic and applied research; for transforming data into informational products tailored to the needs of a variety of user groups; for management of coastal environments and their living resources; for formulating regulations; and for monitoring and assessment. NOAA doesn't have full responsibility for any of these activities, but it is the only agency that I know of that has some responsibility for all of them. The comic character, Pogo, once observed that "some opportunities are so large, they are insurmountable." NOAA's opportunity is very large, but not insurmountable.

On the Need For New Science-Management Paradigms

We need new paradigms for managing coastal systems, at least for those that receive wastes from a diverse and complex array of point and non-point sources throughout much, or all, of each systems drainage basin. In the next presentation, I expect that you will hear about one such paradigm--integrated coastal management. It is an elegant paradigm. How could one quarrel with it? It includes all the politically correct concepts--ecological risk assessment, risk management, integrated management, research, monitoring, feedback, partnerships--all wrapped up--in one neat package called integrated coastal management. I am part of the National Research Council's Water Science and Technology Board's Committee on Wastewater Management for Coastal Urban Areas that has struggled for well over a year with the problems of waste management in urban coastal areas and that produced the draft report on integrated coastal management. We've made some progress, but the hard part is still ahead: to show how to apply the paradigm not in the abstract, but in specific, concrete terms the way managers would have to do it. If those of us who developed the concept do not, or can not, test it, how can we expect others to use it? In his new book Sur/Petition (1992), Edward de Bono states:

A concept that is not important, after all, is a wasted concept. A concept is unlikely to be implemented if it cannot be tested. That is why it is important to design a concept not only for eventual use, but also for preliminary testing. A concept that can show its benefits in preliminary testing stands a better chance of getting used than one which can not.

Another approach is the Estuarine Science-Management Paradigm developed through the Marine Sciences Research Center (1990, 1991). It describes a new model for forging and maintaining partnerships among key decisionmakers, scientists, educators, and public interest groups. Like integrated coastal management, it remains untested. But it now has the endorsement of several key individuals, foundations, and resource management agencies in New York, and we expect to put it to the test over the next year.

A CLOSING OBSERVATION

When the curtain goes up for the opening scene in the late British playwrite, Thomas Shadwell's, play the *Virtuoso*, the main character, Sir Nicholas Jimcrack, is seen making froglike swimming motions on his laboratory table. "Do you intend to try it in the water?" he is asked. He responds: "Never sir, I hate the water." And he adds "I content myself with the speculative part of swimming and care not for the practical. I seldom bring anything to use; it's not my way. Knowledge is my ultimate end."

Too many of us in the coastal marine science community are Sir Nicholas Jimcracks. It's time for more of us to get off our laboratory benches and get out into the water. If we want the results of our basic research to be used in a timely way and more effectively, we need to form partnerships with resource managers and decisionmakers. We need to be responsive to their needs and take more active roles in transforming advances in knowledge into forms that can be used to conserve and, when necessary, to restore our coastal environments. It will not be easy, but partnerships never are.

ACKNOWLEDGEMENTS

I thank Doreen Monteleone, Chongle Zhang, Jiong Shen, Andrew Matthews and Kristen Romans for their assistance in preparing this paper.

REFERENCES

ASLO. 1990. At the Land-Sea Interface: A Call for Basic Research. American Society of Limnology and Oceanography, Estuarine Research Federation, and Southern Association of Marine Laboratories. Washington, DC: Joint Oceanographic Institutions.

Bayha, K. 1981. Overview of freshwater inflow. Vol. 2. Pp. 231-247 in Proceedings of the National Symposium on Freshwater Inflow to Estuaries. Washington, DC: National Technical Information Services.

Beaumont, P. 1978. Man's impact on river systems: a world-wide view. Area 10:38-41.

Bokuniewicz, H. 1990. Towards a Framework for Research in Estuaries: The Report of a Workshop held at the Marine Sciences Research Center (MSRC), State University of New York at Stony Brook, New York. MSRC Working Paper #40, Reference #90-5.

Brink, K., J. Bane, T. Church, C. Fuirall, G. Geernaert, D. Hammond, S. Henrichs, C. Martens, C. Nittrouer, O. Rodgers, M. Roman, J. Roughgarden, R. Smith, L. Wright, and J. Yoder. 1992. Coastal Ocean Processes: A Science Prospectus. Woods Hole Oceanographic Institution, Woods Hole, MA 02543. 88p.

Clark, J., and N. Benson. 1981. Summary and Recommendations of Symposium. Vol 2. Pp. 523-528 in Proceedings of the National Symposium on Freshwater Inflow to Estuaries. Washington, DC: National Technical Information Service.

Copeland, B.J., K. Hart, N. Davis, and S. Friday, eds. 1984. Research for Managing the Nation's Estuaries: Proceedings of a Conference in Raleigh, North Carolina. Sponsored by the National Sea Grant College Program and the National Marine Fisheries Service. UNC Sea Grant College Publication UNC-SG-84-08.

CRC. 1983. Cronin, L.E. ed. 1983. Ten Critical Questions for Chesapeake Bay in Research and Related Matters. Chesapeake Research Consortium, Maryland.

Croome, R. L., P. A. Tyler, K. F. Walker, and W. D. Williams. 1976. A linmological survey of the River Murray in the Albury-Wodonga area. Search 7(1):14-17.

Culliton, et al. 1990. 50 Years of Population Change Along the Nation's Coasts, 1960-2010. Rockville, Maryland: NOAA, Strategic Assessments Branch, Ocean Assessments Division, National Ocean Service.

DeBono, E. 1992. Sur/Petition. New York, New York: Harper Business, a Division of Harper Collins Publishers.

GESAMP. 1991. The State of the Marine Environment. Blackwell Scientific Publications, Oxford Press.

Goldberg. 1990. Protecting the wet commons. Environmental Science and Technology. 24:450-454.

Hammond, D. E., ed., 1974. Recommendations for Basic Research on Transfer Processes in Continental and Coastal Waters: An Essential Ingredient for Predicting the Fate of Energy-Related Pollutants. A Report to the National Science Foundation Submitted by the Participants in a Workshop held at Block Island, Rhode Island. July 23-25, 1974.

Herrgesell, P. L., D. W. Kohlhurst, L. W. Miller, and D. E. Stevens. 1981. Effects of freshwater flow on fishery resources in the Sacramento-San Joaquin estuary. Pp. 71-118 In: Proceedings of the National Symposium on Freshwater Inflow to Estuaries, Vol. 2. Washington, DC: National Technical Information Service.

Kim. 1990. Essence of Creativity. New York and Oxford, England: Oxford University Press.

Limburg, K. E., C. C. Harwell, S. A. Levin, eds. 1984. Principle for Estuarine Impact Assessment: Lessons Learned from the Hudson River and Other Estuarine Experiences. Hudson River Foundation. Ithaca, New York: Cornell University.

Marine Sciences Research Center. 1991. On Development of an Estuarine Science-Management Paradigm. MSRC Working Paper 50, Reference 91-07.

Marine Sciences Research Center. 1990. On Development of an Estuarine Science-Management Paradigm. MSRC Working Paper 46, Reference 90-15.

Meade, R. H., T. R. Yuzyk, and T. J. Day. 1990. Movement and storage of sediment in rivers of the United States and Canada. In Wolman, M.G., and H.C. Riggs, eds., The Geology of North America, vol. 0-1. Surface Water Hydrology. Boulder, Colorado: Geological Society of America.

Merkle, P. B. and B. J. Brownawell. In press. A weather driven fugacity model of the atmospheric deposition of semivolatile organic compounds to aquatic environments. Environmental Science and Technology.

NASULGC. 1986. On the Importance of Estuarine Research. National Association of State Universities and Land Grant Colleges, Marine Division, Estuarine Committee.

NERC. 1982. Research on Estuarine Processes. Report of a Multidisciplinary Workshop held at the University of East Anglia on 14-17 September 1982. National Environment NOAA, 1990. National Status and Trends Program, 1990. Coastal Environmental Quality in the United States, 1990: Chemical Contamination in Sediment and Tissues. 34 p. Research Council.

NOAA. 1991. The 1990 National Shellfish Register of Classified Estuarine Water. National Oceanic and Atmospheric Administration, National Ocean Service, Office of Oceanography and Marine Assessment, Strategic Assessment Branch.

NOAA. 1986. Coastal Marine Research Plan. Draft 11/13/86. National Oceanic and Atmospheric Administration.

NRC. 1983. Fundamental Research on Estuaries: The Importance of an Interdisciplinary Approach. Geophysics Research Board, National Research Council, Washington, DC: National Academy Press.

NRC. 1977. Estuaries, Geophysics, and the Environment. Geophysics Research Board, National Research Council: Washington, DC: National Academy Press.

Rosengurt, M., and I. Haydock. 1981. Methods of computation and ecological regulation of the salinity in regime in estuarine and shallow seas in connection with water regulation for human requirements. Pp. 474-506 in Proceedings of the National Symposium on Freshwater Inflow to Estuaries, vol. 2. Washington, DC: National Technical Information Service.

Schubel, J. R., Y. Eschet, C. Zhang, J. Shen and R. Nino-Lopez. 1991. Human Effects on the Discharges of Water and Sediment by the World's Rivers: An Overview. Stony Brook, New York: Marine Sciences Research Center, State University of New York.

Schubel, J. R., and D. W. Pritchard. 1991. Some Possible Futures of Long Island Sound. Marine Sciences Research Center Working Paper 55, Reference 91-17.

Schubel, J. R., and D. M. Monteleone. 1990. Critical Problems of New York's Marine Coastal Zone: A Preliminary Selection. Marine Sciences Research Center Working Paper No. 44, Ref. No. 90-11. Three Sections.

Simon, H. A. 1986. What we know about the creative process. Chapter 1 (pp. 3-20) in Frontiers in Creative and Innovative Management, R.L. Kuhn, ed. Cambridge, Massachusetts: Ballinger Publishing Co.

Titus, James G. 1991. Greenhouse effect and coastal wetland policy: How Americans could abandon an area the size of Massachusetts. Environmental Management.

Titus, James G. 1990. Greenhouse effect, sea level rise and land use. Land Use Policy 7(2):138-153.

USEPA. 1990. The Quality of Our Nation's Water.

10

Coastal Management and Policy

William Eichbaum
The World Wildlife Fund
Washington, D. C.

This paper is one of several prepared for a retreat on coastal zone issues sponsored by the National Research Council's Commission on Geosciences, Environment, and Resources. A number of papers discuss focused technical aspects of the coastal zone such as ocean circulation and coastal meteorology, and others deal with broader subjects such as coastal wetlands and coastal/nearshore littoral systems, land use and the coastal zone, and coastal pollution and waste management. The topic of this paper, "Coastal Management and Policy," inevitably touches on aspects of each of the others. In an effort to minimize the duplication of discussion, I will specifically address three topics that are generic in nature. The first concerns the question of how we are organized to provide governance of the coast. Second, I will discuss several problems encountered in translating science into policy. Finally, I will touch upon the question of how human values and expectations affects the problem of coastal management. In the conclusion, I will offer a suggestion for improving the development of policy for the coast and its management.

THE GOVERNANCE SYSTEM

Perhaps the most salient feature about coastal governance in the United States is the extraordinary degree to which it is fragmented.[1] This fragmentation exists in two dimensions that have important ramifications.

As a consequence of our federal system of government and a concurrent fierce attachment to local authority, coastal governance is divided among at least three levels--federal, state, and local. (Increasingly, the international agenda adds a significant, additional level of authority for coastal matters.)

---

[1]There are a number of intellectual or policy efforts to overcome this fragmentation. They include such ideas integrated permitting and inspection and sustinable development.

In some instances the governance authority is exclusively held at only one level of government. For example, land-use controls in the interest of flood or erosion management generally are exclusively exercised, if at all, at the local level, while management of navigational systems is largely a federal responsibility.

Other instances are found where authority may be simultaneously and separately exercised by two levels of government. For example, in many states,
activities to regulate development in wetlands are carried out by both the states and the federal government.

There are models of governance that provide for the delegation of authority from one level of government to another. Thus, the Federal Clean Water Act provides for delegating most federal authority to the states, and this has been done in many regions of the country.

In some circumstances, the same resource may be regulated by a different level of government depending upon the geographical region of the coastal zone in which it is located. Many species of fish are regulated pursuant to a federal system when located offshore, but the same species (and individual) may be regulated by state authorities when it migrates to freshwater rivers.

Authority may exist at different levels of government with respect to a particular issue depending upon the function being performed regarding that issue. Thus, the responsibility for constructing domestic sewage treatment plants is generally at the local level, while the responsibility for setting minimum standards and (until recently) raising the necessary funds was a federal one.

In addition, there are any number of ways in which these three levels may choose to combine in order to regulate or manage a particular coastal resource or activity. Two or more states, perhaps with the federal government or local authorities, may join together in an interstate compact to create new authorities to manage a particular resource or activity. This has been common for interstate bodies of water and for regional port facilities. They may also reach more informal arrangements that are designed simply to assure a high degree of coordination of multijurisdictional activities--the Chesapeake Bay Program is a classic example.

The subject matter, or issue, of fragmentation of coastal governance adds further complexity. Responsibility for various substantive issues is always widely distributed among a very large number of government organizations at every level. The extent of this fragmentation is well known and need not be repeated. There are, however, several characteristics that are worth mentioning.

In many cases this dispersal of authority is not crisp. There tends to be a certain gradation in responsibility from one agency to the next. For example, while the Environmental Protection Agency is clearly responsible for non-point- source-water quality issues and also manages the nation's national estuary program, the Department of Commerce, through the National Oceanic and Atmospheric Administration and Coastal Zone Management Act responsibilities, also has non-point source water quality functions, as does the Department of Agriculture, through the Soil Conservation Service. The blurring of authorities and responsibilities as one moves from the core function of one agency to that of another can have the disadvantage of diffusing authority. On the other hand, it can also allow for competition among agencies to do a good job.

Even where governments have attempted consolidation in order to eliminate fragmentation of governance, success has been rare. Very large organizations with a multiplicity of missions tend to be internally diffuse. Furthermore, it is simply not possible to combine everything. Recognizing these short-comings, the tendency during the early 1970s to create *super* agencies seems to have ended. A notable exception is the recent effort of California to create a unified Department of the Environment. Even within that strengthened department, water-quality-management responsibility is geographically divided among a number of very strong regional water boards. And, coastal management per se is located in the separate Resources Department.

The foregoing discussion has largely focused on the program fragmentation within and among the executive branches of governments. It is important to note that, at least at the federal level, the problem of fragmentation within the legislative branch of government is also severe. There are well over thirty subcommittees of Congress responsible for matters relating to the coastal environment.

USING SCIENCE TO INFORM POLICY AND ACTION

While fragmented systems of governance have contributed mightily to the current poor management of U.S. coastal resources, the many problems associated with properly focusing science on the policy choices have been as harmful. The problem of the relationship between science and policy is not unique to the coast and has been extensively treated by a number of authors. The essential question is how to organize scientific information in a fashion that is understandable to policymakers and that compels an effective management response even though the information itself is imperfect. There are some peculiar aspects of the coastal marine environment that make this problem even more severe. They do bear some discussion.

Firstly, most of what goes on in the marine environment is invisible to all but the most sophisticated and dedicated investigator. This invisibility condition makes it almost impossible for the citizen or policymaker to have any intuitive sense of the actual condition of the resource as it may be described by the scientific community. Lacking this, a sense of reality and even urgency, where warranted, is missing. A useful comparison between the effects of invisibility of the marine environment and visibility in the terrestrial environment is found in the Chesapeake Bay. During the 1980s, over 80 percent of the bay's submerged aquatic grasses died. Scarcely anyone noticed. Imagine the public reaction if 80 percent of the forest resources of the bay's watershed had died during the decade. Lack of visibility of marine processes also means that sometimes what is observed can be distorted far beyond the actual importance that the observed event may have scientifically. Thus the washing up of a relatively small number of syringes on the beaches of New Jersey during the summer of 1988 crystallized a public impression that may have been far removed from the scientific reality.

Secondly, the actual science of the coastal environment is extraordinarily complex. Crucial processes take place in the atmosphere, on the land, and within the water. They may be immediate; a watershed away; or, as in the case of El Ninjo, half a world away. They can be physical, chemical, biologic, or some combination of the three. Much about these relationships is not well known. In addition to the extreme complexity of the natural systems, the possible ways in which human activities can intrude are even more complex. They range across the full spectrum of economic and recreational life not just on the coast but throughout the watershed. These intrusions include pollution, to degradation of fisheries, to development in flood plains, to habitat destruction for homes, and to materials used such as pesticides for agriculture. The sum is hugely complicated, and it has only been in the last ten years that society has really begun to think of the total interaction in anything like a science-based systems way. This inevitably means that for some time to come, most questions about the nature and causes of the problems of the marine environment will be answerable in only the most tentative and perhaps inconclusive terms. This poses serious problems for policymakers seeking to allocate scarce resources as well as to the public, which is looking for the certainty of environmental protection.

Finally, the inherent complexity of the marine environment and its relationship to land-based activities makes the problem of cumulative impacts especially severe. All too often, adverse environmental consequences take place as a result of a large number of relatively minor activities that occurred long before the consequences themselves. Often the eventual adverse environmental event is dramatic in nature and nearly irreversible. For example, it is difficult to assess the impact on the coastal environment of the conversion of any individual farm or wood lot to a tract housing complex. Yet, there are numerous instances where widespread conversions across a region have resulted in the general decline of coastal resources and, specifically, in the loss of important shellfish resources.

Just as governance has been fragmented, the practice of the science of the marine environment (as much other environmental science) has been fragmented and myopic. For science to be most effective in the face of uncertain knowledge, it needs to develop information from as wide a range of disciplines as possible and search for the integrating themes that begin to illuminate paths for action. For a variety of reasons, this kind of science is neither valued by the academic community nor often sought by the government.

PUBLIC VALUES AND ROLES

There is no single *public* nor is there a uniform set of values regarding the coastal environment. While there is great diversity, a common driving interest can be seen. That common interest is use. Various parts of the public will tend to define their objective for management in terms of whether the use they make of the coastal environment is protected. Fishermen and seafood consumers will want to be assured that seafood is safe to consume. Surfers and swimmers will want to be certain that it is safe to swim in the water. Those who appreciate the enjoyment of the marine environment, such as bird watchers, will want to know that the ecological system is healthy and viable. And commercial and recreational fisherman will expect that the productive quality of the marine environment is protected.

In summary, the various sectors of the public will desire the following:

- fish and shellfish that are safe to consume;
- water that is safe to come into contact with;

- a healthy ecological system; and
- a productive ecological system.

The foregoing discussion suggests that there is not a simply stated set of public objectives for coastal environments. This diversity is reasonable since divergent objectives depend upon the viewpoint of the particular participant in the discussion. No view can be seen as wrong when analyzed from the stance of its own interest and the societal values that it seeks to advance or protect. However, the very fact of the variety of interests and sources of threats suggests one basis of a complex process of risk analysis and priority setting.

A factor which further complicates this already murky picture is that the mix of these various values changes constantly with time. Values held by particular interests can change, and the relative importance that is ascribed to different values by society can change. In addition, the objective factual setting will evolve. Scientific understanding and technological capacity do grow. Choices about risk will vary. The net result, however, during this century, has been a steadily growing body of knowledge about the extent to which the environment is being damaged and steadily growing public demand for improved levels of environmental protection.

CONCLUSION

There are severe consequences flowing from this fragmented and complex system of governance. Of course, there are the usual problems of waste, duplication of effort, lack of coordination on common problems, and conflicting political agendas. While serious, these are not the most important consequences of the current fragmentation of coastal governance. More important problems are

- collective failure to identify the most important threats to the quality of the coastal environment;
- failure to design a responsive management strategy that allocates scarce resources to the most critical problems;
- occasional massive attention to a high-profile condition that, even when resolved, will still not solve the identified coastal problem;
- an inability of public attention to focus on one political entity as responsible and accountable for the improvement of the coastal environment; and
- a distortion of science in that initially clear, positive research studies result in confused conclusions.

In essence, all this means that (1) unwise actions are often taken, (2) responsibility and accountability for the wise management of coastal resources is diffuse, and (3) the process is often inaccessible to concerned publics.

There have been a limited number of efforts to improve this situation. At the federal level, the only significant attempt was the enactment of the Coastal Zone Management Act in 1972. It can be argued that this statute provided the opportunity for the federal government to initiate a process of forging, at the state level, a system of governance that would meld the disparate functions into a coherent whole. However, this opportunity was generally missed as the federal program only sought to achieve a state coastal program designed around a system of various, loosely coordinating authorities. The result has been that separate authorities remain dominant and that common actions are still lacking.

Solutions to these problems can be found through implementation of a governance strategy based on the concepts of integrated coastal management. In brief, integrated coastal management is a methodology that identifies important scientific and human-value issues on an ecological basis, compares the risks posed, and develops risk management options that effectively allocate scarce resources to the most important problems. This dynamic process is action oriented and iterative with refinements being based on monitoring, research, and institutional responses.

The first step in actually implementing such a strategy would be to identify a single government entity to be assigned the actual responsibility of assuring that integrated coastal management is carried out. This does not mean that all governmental functions must be combined in one *superagency*. In highly complex or geographically widespread situations, it may mean that new coordinating bodies need to be established. In either case, in order to be effective the responsible entity should have the authority in the following areas: planning for integrated coastal management, monitoring for environmental results, coordinating of budgets, and data management.

11

Research and Development Funding for Coastal Science and Management in the United States

R. Eugene Turner
Louisiana State Unviersity
Baton Rouge, Louisiana

Jerry R. Schubel
The State Unviersity of New York
Stony Brook, New York

INTRODUCTION

It is an indisputable conclusion that coastal ecosystems are important to society, yet many are stressed and require remedial action, and conserving the remaining values is unpredictable. It is not surprising, therefore, that various strategies exist to achieve the desired goals for these systems. However, after decades of much work and financing, these goals are still elusive in the case of most coastal systems. While old pressures continue, new ones arise (e.g., aquaculture, eutrophication, and sea-level rise). Yet the natural attributes and *controlling circuitry* are still incompletely understood. Two activities are necessary: the development of new knowledge and the application of that knowledge.

Both scientists and engineers (S&E) are involved in these two activities. The purpose of preparing this material is to examine important funding patterns and cross-sectoral interactions among government, education, and industry. This is done to help evaluate and devise recommendations affecting how we (managers, administrators, scientists, and engineers) may more usefully contribute professionally. We used mostly national statistical parameters because coastal scientists and engineers are obviously not a distinctive tribe professionally isolated from either their inland peers or offshore colleagues.

WHERE ARE THE COASTAL SCIENTISTS AND ENGINEERS?

Coastal S&E cannot be defined as easily as chemists or physicists, which are from more traditional disciplinary interests. Coastal fields tend to be interdisciplinary, may include management, and are limited to the coastal zone (which is itself undefined). Traditional societal surveys may therefore not recognize all significant participants. An American Geophysical Union (1986) survey of ocean scientists and engineers was used here as an indicator of the total number of coastal S&E and as a descriptor of their geographic distribution.

In 1986 there were 6000 people included in the American Geophysical Union (AGU) survey, which compares to the national S&E manpower of 400,358 (NSB, 1991). This is 1.5 percent of the national S&E workforce. Eighty-five percent of those surveyed were in the coastal states, whose geographic distribution is described in Figure 11.1. There are significant numbers of coastal S&E in all coastal states. The density per population is highest in the northeast and northwest coastal sectors (upper panel), and density per shoreline length (from the World Almanac) is highest in Maine, Alaska, Gulf of Mexico, and the southeastern United States, (middle panel, Figure 11.1). There are 11 institutions with more than 50 S&E among the 25 coastal states (lower panel). There is a ten-fold range from minimum to maximum in all values. Obvious centers of concentration are spread broadly within all regions of the United States. All states appear to have a minimal core of S&E working in the coastal zone, however it is defined.

NATIONAL SCIENCE AND ENGINEERING ISSUES

Coastal scientists and engineers are less than 10 percent of the total U.S. S&E manpower, and their sectoral addresses (per above) are not obviously different from the rest of the national S&E workforce. National statistics may provide information, therefore, about the activities and resources of coastal S&E.

National indicators of science and engineering funding and manpower have increased tremendously since World War II. In 1940 there were 330 Ph.D.s per one million people older than 22 years. By 1966 that ratio climbed to 778:1,000,000, in 1970 it was 1587:1,000,000, in 1990 it was 2000:1,000,000 (Stephan and Levin, 1991). In 1976, S&E employment was 2.4 percent of the workforce, but in 1986 it was 3.6 percent. The number of United States baccalaureates and first professional degrees has increased about 4.8 percent annually since 1900. The new workforce is being trained today but apparently not in sufficient quantity to meet modest projections for national needs. Various reviews (e.g., Pool, 1990; Atkinson, 1990; NSB, 1991) suggest an "annual supply-demand gap of several thousand scientists and engineers at the Ph.D. level, with the shortage persisting well into the twenty-first century" (Atkinson, 1990) as an aging faculty retires, as the student population bulges, and if historical S&E employment growth continues. Employment growth for S&E, however, is not projected to remain steady but to rise.

What is the support for the present and future graduate students? Where will we find those young S&E for work in the coastal zone?

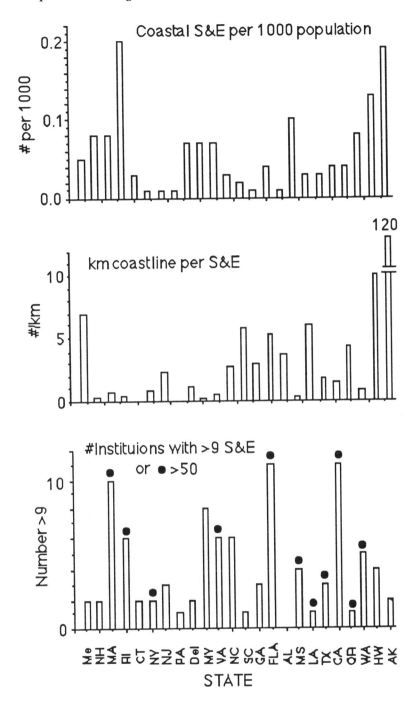

FIGURE 11.1 Geographic distribution of ocean scientists and engineers in the U.S. coastal states. Top: coastal S&E per state population (1990 census data). Middle: Coastal S&E per km tidal coastline. Bottom: Number of coastal S&E per state. The height of the bar indicates more than 9 reside at one address. A "•" indicates more than 50 reside at one address (maximum of one in any one state). Data are from a 1986 survey of scientists and engineers (American Geophysical Union, 1986). Reprinted with permission from American Geophysical Union, 1986.

In 1969 there were 80,000 S&E graduate students supported by federal funds (Table 11.1). In 1980 and 1990 the federal support changed from 44,590 to 52,875 students (an 18.5 percent rise), respectively, and 20 percent of the total were supported on federal funds. Meanwhile, the student population rose to more than 267,621 students and the S&E workforce increased 50 percent. The number of environmental students (1-1.5 percent of the total) has declined in the past 10 years. From 1980 to 1990, total S&E expenditures increased 130 percent in current dollars, and 51 percent in constant 1982 dollars. Those dollars not only paid for basic salaries of a more experienced workforce but for increasingly sophisticated equipment amidst the growth of larger projects. In other words, in the past decade: (1) the dollars for Research and Development (R&D) and the S&E workforce rose together, (2) student support rose slower than total support, and (3) the average project size per potential investigator went down. One consequence is that education of graduate students may be compromised. One cannot (or should not) turn students loose without supervision, and they need resources to be trained.

The experience at the National Science Foundation (NSF) reflects these changes. The proposal success rate at NSF has gone from 38.5 percent in 1981 to around 30 percent in 1990 (Palca, 1990). The average grant size has dropped from $68 thousand in 1985 to $61.7 thousand in 1989 (1989 dollars). Individual investigators accounted for 57 percent of the research budget in 1991 compared to 68 percent in 1980. Although centers are often blamed for the lack of small science, only 4.5 percent of the NSF research budgets go to research centers. The change in average grant size has not gone unnoticed. Dalrymple (1991) quoted the V. Bush report (1945) setting up the NSF: "New products and processes are founded on new principles and conceptions which, in turn, are developed by research in the purest realms of science." The concern is twofold: (1) that directed science compromises undirected science and (2) that big, multi-investigator projects compromise the productivity of the smaller, often single-investigator projects. The resolution of these issues affects how coastal zone science and management will proceed. There are urgent pleas to have scientists actively involved in the process of setting priorities (e.g., Kaarsberg and Park, 1991) for a shrinking budget.

However, even if funding increases in the near future, it is unlikely that we will return to the days immediately following World War II, when it was said that "under present conditions, the ceiling on R&D activities is fixed by the availability of trained personnel, rather than by the amounts of money available" (Steelman, 1947). Unfortunately, both funding and manpower appear limiting in the years ahead, and the S&E community should prepare for strenuous discussions about the relative merits of big and small science, and of directed vs. undirected research. In the midst of these changes appears the complicating (and compromising) growth of *ear-marked* funds for R&D, otherwise known as *pork barrel* projects. In 1983 there were three of these projects worth about $16 million (Figure 11.2). By 1990 the amount rose to nearly $500 million. An Office of Science and Technology Policy analysis for 1991 appropriations bills identified 492 projects totaling $810 million. For comparison, this is equal to 44 percent of the NSF budget, and 6 percent of the national educational R&D budget. Scientists and engineers should be wary of this development, as it is a substitute for merit review. Merit review must be effective and acceptable to be a plausible defense against pork barrel funding.

TABLE 11.1 Changes in Support for S&E and Graduate Students in 1969, 1980, and 1990. Data are in 1992 dollars.

	1969	1980	1990
Graduate S&E Students to Education			
Federal Sources (no.)	80,000	44,590	52,875
All Sources (no.)	-	215,354	267,621
% Federal	-	20.7%	19.7%
Environmental Studies (no.)	-	3,442	2,939
% Environmental Students	-	1.6%	1.1%
Research and Development to Education			
Federal (billion $)	1.6	4.1	9.3
All Sources (billion $)	2.2	6.1	16.0
% Federal	71.9%	67%	58%

Adapted from data in Atkinson, 1990 and NSB, 1991.

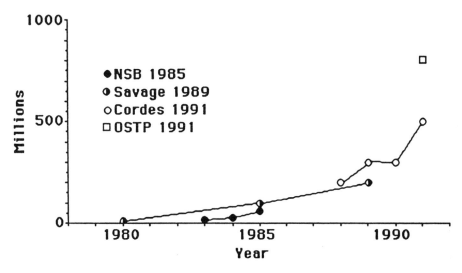

FIGURE 11.2 The rise in non-competitive, or *earmarked* funds disbursed outside agency initiated requests. Adapted from data reported in NSB, 1991. Note: the 1991 NSF budget was $1.954 billion, and the total research and development funds for educational institutions in 1988 - 1989 was $13.5 billion.

RESEARCH PRODUCT QUALITY

There are three major performers of research: government, industry and educational institutions. Government laboratories have a reputation for focusing on project management, monitoring and immediate problems over publication; educational laboratories for a *publish-or-perish* reputation; and industry for garnering funds necessary for either profit or proprietary interests and for a rapid response and on-time performance, etc. These three sectors have legitimate viewpoints and functions that a national research and development policy should be matched with, goal for function.

The educational community clearly excels at developing new information. Two demonstrations of this are discussed here. Analyses of frequently cited papers in ecology and oceanography are provided by McIntosh (1989) and Garfield (1987), respectively. The most frequently cited papers, indicators of highly useful scientific contributions, are dominated by authors residing or working with educational institutions (Table 11.2). More than 95 percent of all *classic* papers are from these institutions. Officer et al. (1981) provided a different type of analyses for selected estuarine publications. They concluded that "the academic community has produced most (7 percent) of the referred research literature on estuaries--evidence of the importance of academic sources of new knowledge" on somewhere between 31--37 percent of the available funding. A more complete quantification of 1200 S&E publications for 1984 (NSB, 1987) indicates that 61 percent of all articles arose from educational institutions (Table 11.3; article number was proportioned according to all author's addresses). This educational contribution was done on 9 percent of the funding and was 14 percent of the average dollar spent per article generated. In other words, the quality and quantity of educational research publications compares very well with all other sectors.

What do other sectors prefer in terms of working with each other on new information? The overwhelming preference is for collaboration with coauthors at educational institutions (Table 11.4). The strongest preference across sectors was for federal scientists to work with scientists in the educational sector. Nearly half of all articles from federal laboratories were co-authored with educational sector coauthors.

TERMINUS

Coastal science and management are not particularly overwhelmed with useful data and more data, are required to address the newly arising complications of increased population, limited resources, and complex management milieus. New scientific and engineering contributions are heavily weighted toward educational R&D contributions. The present R&D funding environment is stagnating. It is threatened by looming manpower shortages and is very competitive. These factors are beginning to place a strain on the S&E community, which is becoming increasingly vocal about the instability of funding for individuals, the size of individual project funding, and the distribution of funding toward fewer and larger projects. Some see fields that are becoming "overcrowded with *risk avoiders* more worried about their next grant" (Stephan and Levin, 1991). The federal government is becoming less involved in R&D in terms of the percent funding, publication, and student support.

TABLE 11.2 Addresses of Authors of *classic* Journal Papers in Ecology and Oceanography. WH and SIO are scientists whose mailing address is Woods Hole, Massachusetts and Scripps Institute of Oceanography, California, respectively. Scientists living there may, or may not, have an academic affiliation at the time of the article publication.

	United States				Foreign	
	WH/SIO	Educ.	Govt.	Industry	Educ.	Govt.
McIntosh, 1989	5	61	2	1	3	26
Garfield, 1987 non-core journals	22	4	0	0	2	1
Garfield, 1987 core journals	10	5	2	0	7	5
Totals	37	70	4	1	41	12

TABLE 11.3 Expenditures (Millions) and Cost Per Article Published by Sectors for 1984. Adapted from data in NSB, 1991, 1987.

Sector	R&D $ (Millions)	% funding	# articles	% articles	$1000/ article
University	8,617	9	30,988	61	278
Non-profit	3,000	3	5,803	11	517
FFRDC*	3,150	3	1,970	4	1,599
Federal	11,572	11	8,898	18	1,301
Industry	74,800	74	2,930	6	25,529
Total	101,139	100	50,599	100	1,999

*FFRDC - Federally Funded Research and Development Center.

TABLE 11.4   Percent Cross Sectoral (education, industry, non-profit, federally financed research and development centers, and federal) Authorship of Journal Articles Published in 1984. The items in **bold** are the two most frequent cross-sectoral associations for the authors in that sector. Adapted from data in NSB, 1987.

	Primary Authors				
	Educational	Industry	Non-profit	FFRDC	Federal
Co-Authors					
Educational	**77**	**24**	**53**	**37**	**48**
Industry	3	**64**	3	6	4
Non-profit	7	3	**36**	2	5
FFRDC	2	3	1	**49**	2
Federal	**10**	6	8	5	**42**

It has been suggested elsewhere that it is appropriate for student support from federal sources to be doubled (Vaughn, 1989). Doing that without depleting the other R&D resources would have the long-term effect of increasing the labor pool in future years and supporting higher education, which does significant amounts of the research.

Partnerships with the educational R&D sectors should be encouraged. The R&D activity of the educational sector is of high quality, high quantity, and is relatively inexpensive compared with other sectors. Federal-educational linkages are pretty good but not with all sectors. We must be careful when co-mingling the different institutions so as not to compromise the qualities of each by confusing their functions, which are not the same. University scientists are not natural resource managers but teachers, even scholars, and technically astute. The political demands of government service require skills that are not taught on class field trips or in the research laboratory. Time spent by government on management and administration is time not available to develop, assimilate, and synthesize new information. Industrial R&D activities are more sensitive to profit, etc. What we do not know is how much more overlap of activities is desirable. The partnerships of the next few years should be interesting in that regard, and attempted gradually.

There are some conspicuous differences between research in the coastal zone and elsewhere that should be mentioned. First, research in the coastal zone is more parochial, in some ways, than in blue water oceanography, for example. In forest research, there is an open competition for funds that is sometimes lacking in coastal- zone research. Also, the role of the research community in proposal review by the former has a greater role in determining quality than the estuarine research. Socio-political aspects tend to be more influential in deciding what questions should be asked and how to manage an estuarine project. Although the pressure is somewhat understandable given the multiple and often conflicting views of estuarine management, the result is often *quick and dirty* R&D projects and an undue influence of opinion about sometimes very sophisticated scientific and technical issues. Good analyses can lay out the options and their implications without involving a policy choice; good policy decisions cannot be made without good analyses. A recent example of the effect of an absence of good analysis is the debacle over redefining wetlands in the absence of scientific judgement (Kusler, 1992, Sipple, 1992). In this example, scientific judgements were phased out in favor of policy outcomes suiting an exclusively political agenda. Second, it is worth repeating that good analyses do not usually come quickly, and that simple plans of actions are usually just that--simplistically inappropriate. The short attention spans of managers and politicians amidst S&E untrained in policy makes for a treacherous liaison between what is an artificial division of 'basic' and *applied* R&D. It is a particularly daunting challenge to derive the essentially interdisciplinary programmatic thrusts necessary to answer the management questions within the coastal zone. Third, unstable financial resources will not be as effective as long-term support. Excellence requires stability (but not entrenchment). Meeting the immediate needs of management compromises achievement of substantial gains over the long haul.

REFERENCES

American Geophysical Union. 1986. U.S. Ocean Scientists and Engineers. Washington, D.C.: American Geophysical Union.

Atkinson, R. C. 1990. Supply and demand for scientists and engineers: A national crisis in the making. Science 248:425-432.

Bush, V. 1945. Science, The Endless Frontier: A Report to the President on a Program for Postwar Scientific Research. Washington, D.C.: National Science Foundation.

Dalrymple, G. B. 1991. The importance of 'small' science. Earth Observation System (EOS). Trans. American Geophysical Union 72:1, 4.

Garfield, E. 1987. Which oceanography journals make the biggest waves? Current Contents 48:3-11.

Kaarsberg, T. M., and R. L. Park. 1991. Scientists must face the unpleasant task of setting priorities. Chronicle of Higher Education. Feb. 20, 1991, A52.

Kusler, J. 1992. Wetlands delineation: An issue of science or politics? Environment 34:7-11, 29-37.

McIntosh, R. P. 1989. Citation classics of ecology. Quarterly Review of Biology 64:31-49.

National Science Board (NSB). 1987. Science and Engineering Indicators--1987. Washington, D.C.: U.S. Government Printing Office.

National Science Board (NSB). 1991. Science and Engineering Indicators--1991. Washington, D.C.: U.S. Government Printing Office.

Officer, C. B., L. E. Cronin, R. B. Biggs, and J. H. Rhyther. 1981. A perspective on estuarine and coastal research funding. Environmental Science and Technology 15:1282-1285.

Palca, J. 1990. NSF: Hard times amid plenty. Science 248:541-543.

Pool, R. 1990. Who will do science in the 1990s? Science 248:433-435.

Sipple, W. S. 1992. Time to move on. National Wetlands Newsletter 14:4-6.

Steelman, J. R. 1947. Science and Public Policy. Washington, D.C.: U.S. Government Printing Office.

Stephan, P., and S. G. Levin. 1991. Research productivity over the life cycle: Evidence for American scientists. American Economic Review 81:114-132.

Vaughn, J. 1989. The federal role in doctoral education. A policy statement of the American Association of Universities, Washington, DC, September, 1989.

Appendix A

Biographical Sketches of Principal Authors

ALAN F. BLUMBERG is a principal scientist and leader of the hydrodynamics group at HydroQual, Inc. He received his B.S. in physics from Fairleigh Dickinson University, an M.A. and Ph.D. in physical oceanography from the Johns Hopkins University, and postdoctoral training at Princeton University. His areas of interest are in estuarine and coastal ocean hydrodynamic modeling, the analysis and interpretation of physical oceanographic data and the development of transport predictions for problems associated with a host of water quality issues. He has published over 75 journal articles and reports in these areas. Dr. Blumberg is very active with the American Society of Civil Engineers and is currently chairman of the Computational Hydraulics Committee.

RICHARD F. BOPP is Associate Professor of the Department of Earth and Environmental Sciences of the Rensselaer Polytechnic Institute. He received his B.S. in Chemistry from MIT and his Ph.D. in Geological Sciences from Columbia University. As a member of the research staff at the Lamont-Doherty Geological Observatory, he worked primarily on the Hudson River and nearby estuaries and coastal systems. His research focused on the development of contaminant chronologies, the geochemistry of organic pollutants and nutrient cycling. From 1990-91, he served as Science Officer on the Hudson River PCB Project at the New York State Department of Environmental Conservation.

WILLIAM M. EICHBAUM received his B.A. from Dartmouth College, and his LL.B. from Harvard Law School. He specializes in environmental law and public policy and is currently a Vice President of International Environmental Quality of the World Wildlife Fund in Washington, DC. Prior to his work there, he held posts, including Undersecretary, Executive Office of Environmental Affairs, Commonwealth of Massachusetts, and Assistant Secretary for Environmental Programs at the Maryland Department of Health and Mental Hygiene. Mr. Eichbaum is a member of the Chesapeake Critical Area Commission, the National Environmental Enforcement Council of the Department of Justice, the Coastal Seas Governance Project, the Patuxent River Commission, and the Environmental Law Institute. He was a member of the National Research Council Committee on Institutional Considerations in Reducing the Generation of Hazardous Industrial Wastes, and Committee on Marine Environmental Monitoring. Currently he serves on the NRC Committee on Wastewater Management for Coastal Urban Areas.

DOUGLAS L. INMAN is Professor of Oceanography and founding Director of the Center for Coastal Studies, Scripps Institution of Oceanography, University of California, San Diego. During his more than thirty years of teaching at Scripps and research in many areas of the world, he has pioneered the field of beach and nearshore processes. He is a Guggenheim Fellow, and he has served as a UNESCO Lecturer in Marine Science in a number of countries. Dr. Inman is the author of over one hundred scientific publications, was technical director for the Orbit Award winning film, "The Beach: A River of Sand", and has received the American Society of Civil Engineers "International Coastal Engineering Award" (1988) and the "Ocean Science Educator Award" (1990) from the Office of Naval Research.

STEPHEN P. LEATHERMAN is the director of the Laboratory for Coastal Research and professor of geomorphology in the Department of Geography at the University of Maryland, College Park. He received his B.S. in geoscience from North Carolina State University and a Ph.D. in environmental sciences form the University of Virginia. His principal research interests are in quantitative coastal geomorphology, coastal geology and hydraulics, and coastal resources management. He has authored and edited 8 books and published over 100 journal articles and reports on storm-generated beach processes, barrier islands dynamics, and sea level risk impacts on coastal areas. Dr. Leatherman was an author of the 1987 National Research Council report on "Responding to Changing Sea Level: Engineering Implications" and served on the NRC Committee on Coastal Erosion Zone Management.

RICHARD ROTUNNO holds a Ph.D. in geophysical fluid dynamics from Princeton University. Since 1980, he has been at the National Center for Atmospheric Research, where he is presently a Senior Scientist. His research interests include: The fluid dynamics of tornadoes and tornado-bearing thunderstorms, squall line thunderstorms, tropical cyclones, the sea breeze, orographically modified flow, and fronts and cyclones. He was the chair of the National Research Council's Panel on Coastal Meteorology.

JERRY R. SCHUBEL holds a B.S. from Alma College, a M.A.T. from Harvard University, and a Ph.D. in oceanography from Johns Hopkins University. His areas of research include estuarine and shallow water sedimentation, suspended sediment transport, interactions of sediment and organisms, pollution effects, continental shelf sedimentation, marine geophysics, and thermal ecology. Currently, he is the Director of the Marine Sciences Research Center, and Dean and Leading Professor of Marine Sciences at SUNY Stony Brook. He also is a State University of New York Distinguished Service Professor. He was the senior editor of Coastal Ocean Pollution Assessment, and chairman of the Outer Continental Shelf Science Committee for the Department of Interior Mineral Management Service. He is a member of the New York Academy of Sciences, and a past president of the Estuarine Research Federation. He was a member of the National Research Council's Committee on Marine Environmental Monitoring. He is currently a member of the Marine Board and serves on the Committee on Wastewater Management for Coastal Urban Areas.

R. EUGENE TURNER is Professor, Coastal Ecology Institute, and, Department of Oceanography and Coastal Sciences, Louisiana State University. He received a B.A. degree from Monmouth College (Ill.), a M.S. degree from Drake University (Biology), and a Ph.D. from the University of Georgia (Ecology). His principal research interests are in quantitative coastal ecology and biological oceanography. He is Chairman, Intecol Wetlands Working Group, serves on several national committees, and is active in scientific matters concerning scientific aspects of coastal environmental management. He is presently a member of the NAS/NRC Marine Board, Habitat Committee, Co-Chair of the EPA Gulf of Mexico Habitat Committee, Book Board of AGU, and Editorial Board Wetlands Ecology and Management.

JOY B. ZEDLER holds a Ph.D. in botany (plant ecology) from the University of Wisconsin. Since 1969 she has been at San Diego State University (SDSU) and is currently a professor of biology at SDSU and director of the Pacific Estuarine Research Laboratory. Her research interests include salt marsh ecology; structure and functioning of coastal wetlands; restoration and construction of wetland ecosystems; effects of rare, extreme events on estuarine ecosystems; dynamics of nutrients and algae in coastal wetlands; and the use of scientific information in the management of coastal habitats. She recently worked on a compilation of literature on the creation and restoration of wetlands for the U.S. Environmental Protection Agency. Dr. Zedler was appointed as a member of the Water Science and Technology Board July 1991.

Appendix B

List of Attendees

David Aubrey
Associate Scientist
Woods Hole Oceanographic Institution
Woods Hole, MA 02543

Robert C. Beardsley
Senior Scientist and Chairman
Department of Physical Oceanography
Woods Hole Oceanographic Institution
Clark 3
Woods Hole, MA 02534

Rosina Bierbaum
Oceans and Environment Program
Office of Technology Assessment
Washington, D.C. 20510-8025

Alan Blumberg
Hydroqual, Inc.
1 Lethbridge Plaza
Mahwah, NJ 07430

Donald Boesch, Director
Horn Point Laboratories
The University of Maryland
P.O. Box 775
Cambridge, MD 21613-0775

Charles Bookman, Director
National Research Council
Marine Board - Room HA 250
2101 Constitution Avenue, N.W.
Washington, D.C. 20418

Richard F. Bopp
Department of Geology
Rensselaer Polytechnic Institute
Troy, NY 12180-3590

James M. Broadus, III
Marine Policy Center
Woods Hole Oceanographic Institution
Woods Hole, MA 02543

Brad Butman, Chief
U.S. Geological Survey
Branch of Atlantic Marine Geology
Woods Hole, MA 02543

Nancy R. Connery
244 Old Stage Road
Woolwich, ME 04579

Craig Cox
Senior Staff Officer
National Research Council
Board on Agriculture
2101 Constitution Avenue, N.W.
 Room HA 394
Washington, D.C. 20418

Paul K. Dayton
Scripps Institution of Oceanography
1240 Ritter Hall (MLRG, A-001)
8602 LaJolla Shores Dr.
La Jolla, CA 02093-0201

Cory Dean
The New York Times
229 West 43 Street
New York, NY 10036

Craig E. Dorman, Director
Woods Hole Oceanographic Institution
Woods Hole, MA 02543

William Eichbaum
Vice President
International Environmental Quality
World Wildlife Fund
1250 Twenty-Fourth Street, NW
Washington, D.C. 20037

John W. Farrington
Associate Director for Education
Dean of Graduate Studies
Woods Hole Oceanographic Institution
Woods Hole, MA 02543

Edward D. Goldberg
Professor of Chemistry
Scripps Institution of Oceanography
University of California, San Diego
La Jolla, CA 92093

John E. Hobbie, Director
Ecosystems Center
Marine Biology Laboratory
Woods Hole, MA 02543

Charles Hollister
Woods Hole Oceanographic Institution
Woods Hole, MA 02543

Douglas Inman
Scripps Institution of Oceanography
University of California
La Jolla, CA 92093

Mary Hope Katsouros, Director
National Research Council
Ocean Studies Board
2101 Constitution Avenue, N.W.
 Room HA 550
Washington, D.C. 20418

Gary Krauss
Senior Staff Officer
National Research Council
Water Science and Technology Board
2101 Constitution Avenue, N.W.
 Room HA 462
Washington, D.C. 20418

Stephen P. Leatherman
University of Maryland
Laboratory for Coastal Research
1113 LeFrak Hall
College Park, Maryland 20742

Gene E. Likens
Director
The New York Botanical Garden
Institute of Ecosystem Studies
Box AB
Millbrook, NY 12545

Syukuro Manabe
Geophysical Fluid Dynamics
  Laboratory
NOAA
Princeton University
P.O. Box 308
Princeton, NJ 08540

Curt Mason
NOAA Coastal Ocean Program Office
Universal Building, Room 518
1825 Connecticut Avenue, N.W.
Washington, D.C. 20235

Marian Mlay
Director
Office of Ocean and Coastal Protection
USEPA
401 M Street, S.W.
Mail Code WH 556-F
Washington, D.C. 20460

Peter Myers
Director
National Research Council
Board on Radioactive Waste Management
2101 Constitution Avenue, N.W.
  Room HA 456
Washington, D.C. 20418

Jack E. Oliver
Irving Porter Church Professor
Department of Geological Sciences
3120 Snee Hall
Cornell University
Ithaca, NY 14853

Curt Olsen
Department of Energy
Office of Health & Environmental
  Research
Office of Energy Research, ER-75
Washington, D.C. 20545

Frank L. Parker
Professor of Civil and
  Environmental Engineering
Department of Civil and
  Environmental Engineering
Box 1596, Station B
Vanderbilt University/Clemson
  University
Nashville, TN 37235

Stephen Parker
Associate Executive Director
National Research Council
Commission on Geosciences, Environment,
  and Resources
2101 Constitution Avenue, N.W.
  Room HA 466
Washington, D.C. 20418

Duncan T. Patten
Director
Center for Environmental Studies
Arizona State University
Tempe, AZ 85287

John Perry
Director
National Research Council
Board on Global Change
2101 Constitution Avenue, N.W.
  Room HA 594
Washington, D.C. 20418

Carlita Perry
Administrative Associate
National Research Council
Commission on Geosciences, Environment,
  and Resources
2101 Constitution Avenue, N.W.
  Room HA 466
Washington, D.C. 20418

David Policansky
Associate Director
National Research Council
Board on Environmental Studies
  and Toxicology
2101 Constitution Avenue, N.W.
  Room HA 354
Washington, D.C. 20418

Stephen Rattien
Executive Director
National Research Council
Commission on Geosciences, Environment,
  and Resources
2101 Constitution Avenue, N.W.
  Room HA 466
Washington, D.C. 20418

Richard Rotunno
National Center for Atmospheric Research
P.O. Box 3000
Boulder, CO 80307-3000

Jerry R. Schubel
Marine Science Research Center
The University at Stony Brook
Stony Brook, NY 11794-5000

Larry L. Smarr
Director, National Center
  for Supercomputing Applications
University of Illinois at
  Urbana-Champaign
151 Computing Applications Bldg.
605 E. Springfield
Champaign, IL 61820

Jeanette Spoon
Administrative Officer
National Research Council
Commission on Geosciences, Environment,
  and Resources
2101 Constitution Avenue, N.W.
  Room HA 466
Washington, D.C. 20418

William Sprigg
Acting Director
National Research Council
Board on Atmospheric Sciences
  and Climate
2101 Constitution Avenue, N.W.
  Room HA 594
Washington, D.C. 20418

Steven M. Stanley
Department of Earth and Planetary
  Sciences
The Johns Hopkins University
Baltimore, MD 21218

Karl K. Turekian
Silliman Professor of Geology
  and Geophysics
Yale University
Box 6666
New Haven, CT 06511

R. Eugene Turner
Chairman
Center for Wetland Resources
Louisiana State University
Baton Rouge, LA 70803

Irvin L. White
Senior Director
Laboratory Energy Programs
Battelle Pacific Northwest
  Laboratories
901 D St., S.W.
Suite 900
Washington, D.C. 20024

M. Gordon Wolman
Professor, Department of Geography
  and Environmental Engineering
313 Ames Hall
The Johns Hopkins University
Baltimore, MD 21218

Arch Wood
Executive Director
National Research Council
Commission on Engineering and Technical
  Systems
2101 Constitution Avenue, N.W.
  Room HA 280
Washington, D.C. 20418

Joy B. Zedler
Biology Department
San Diego State University
San Diego, California 92182-0057

[illegible handwritten notes]